Biomedical Therapies for Fatal Diseases

Biomedical Therapies for Fatal Diseases documents the history of treating various fatal diseases and the progress made on various therapies that have extended life expectancy and alleviated suffering in those afflicted by these diseases. For example, advances in vaccination protocols have rendered some diseases, which were once almost always fatal, nearly extinct. After the introductory explanation, this book explains causes or mechanisms of fatal diseases such as cancer and ALS, and recent therapies that may relieve patients' suffering, such as photoimmunotherapy, checkpoint blockade therapy, CAR T-cell therapy, and current ALS therapy. Furthermore, clinical trials approved by the FDA are described.

Tadayoshi Shioyama received his bachelor's degree in physics in 1966 and doctorate degree in engineering in 1984 from Kyoto University's Department of Mathematics and Applied Physics. He was inaugurated as professor at the Kyoto Institute of Technology in 1992. Since 2007, he has been professor emeritus at the Kyoto Institute of Technology.

Biomedical Therapies for Fatal Diseases

Tadayoshi Shioyama

CRC Press
Taylor & Francis Group
Boca Raton London New York

CRC Press is an imprint of the
Taylor & Francis Group, an **informa** business

Designed cover image: Shutterstock

First edition published 2027
by CRC Press
2385 NW Executive Center Drive, Suite 320, Boca Raton FL 33431

and by CRC Press
4 Park Square, Milton Park, Abingdon, Oxon, OX14 4RN

CRC Press is an imprint of Taylor & Francis Group, LLC

ISBN: 978-1-041-31771-5 (hbk)
ISBN: 978-1-041-31772-2 (pbk)
ISBN: 978-1-003-78066-3 (ebk)

DOI: 10.1201/9781003780663

Typeset in Times
by Apex CoVantage, LLC

Contents

Preface

In industrialized countries, one of the three leading causes of death is cancer, which is likely to become the most common fatal disease. The others are infectious disease and cardiovascular disease, treatments for which continue to be improved, increasing the average life expectancy. Curing cancer requires that all the malignant cells are removed or destroyed without killing the patient, successfully inducing the immune response against the tumor to discriminate between the tumor and their normal-cell counterparts. In conventional cancer therapies, this discrimination is not complete, leading to damage of normal cells, that is to say, side effects.

Furthermore, ALS (amyotrophic lateral sclerosis) is an incurable motor neuron disease (MND) characterized by degenerative changes in motor neurons. Onset presents as relentlessly progressive muscle atrophy and weakness which influences respiratory muscles, limiting survival to 2–4 years after disease onset in most cases, with the etiology remaining unexplained in over 90% of patients.

During treatment of cancer, patients are suffering from anti-cancer drugs' side effects, such as serious disease derived from drugs (for example, interstitial pneumonia), nausea, little appetite, baldness, swelling, tiredness, and so on, due to damage of normal cells by the drug. Unless there is a therapy completely curing cancer, patients are always anxious about their health, because cancer cells are not completely destroyed by conventional therapy. Such suffering of patients will never be dissolved, unless a therapy that completely cures cancer is developed. Since in sALS cases (which are the 90–95% of ALS cases) (see Chapter 9) the etiology has remained unexplained, most patients with ALS have become pessimistic about their future.

In order to answer the questions of how the human body protects against disability of health, and how human healthcare has been held by biomedical treatments for the diseases, the author has researched biomedical references.

The historical background is as follows. During the Peloponnesian War more than 2,000 years ago, two successive plague infections occurred in Athens. It was noted by Greek historian Thucydides that people who survived the first infection did not contract the disease at the second infection.

As a scientific discipline, the beginning of immunology is attributed to the work of E. Jenner in the late eighteenth century. In 1796, he demonstrated that inoculation with cowpox protected the recipient against smallpox. E. Jenner called this process vaccination. E. Jenner's work was extended by many great microbiologists' discoveries in the late nineteenth century. R. Koch proved that infectious disease is caused by specific microorganisms. In the 1880s, L. Pasteur devised a rabies vaccine. In the early 1890s, E. von Behring and S. Kitasato discovered that the serum of animals with immunity against diphtheria contain a specific "antitoxic activity" capable of conferring short-lived protection against diphtheria in people. The activity was later determined to be due to the protein called an "antibody."

In 1909, Nobel prize winner P. Ehrlich proposed for the first time that the immune system could be used to treat cancer, suggesting that antibodies could be used to deliver toxins to cancer cells. He researched chemotherapy. Since the twentieth century, immunology has been researched on the basis of molecular biology. It has been revealed how the human body's innate immune system defends it by an early immune response against pathogens and how the adaptive immune system strictly protects the human body by antigen-specific recognition against encountered pathogens, killing APCs (antigen-presenting cells) and making immunological memory. Although immunological approaches have treated cancer for more than a century, it is only in the past decade that real promise has been shown.

Because conventional cancer therapies have burdened patients with side effects, it was demanded to develop a cancer therapy without side effects. A new therapy for removal specific to cancer cells without damaging normal cells (without side effects) was developed by H. Kobayashi in the NIH (National Institutes of Health), using antibody-photoabsorbing chemical conjugate specifically targeting tumors, killing cancer cells by illuminating them with near-infrared light. The therapy was called near-infrared photoimmunotherapy (NIR-PIT). In the US, phase II clinical trials of NIR-PIT ended in 2018.

The mechanism of cancer's escape from immunity was probed by T. Honjo and J.P. Allison who independently developed new immunological therapies with a checkpoint blockade, inhibiting cancer's escape from immunity. The immune checkpoint blockade therapy received FDA (Food and Drug Administration) approval. In 2018, T. Honjo and J.P. Allison were awarded the Nobel Prize in Physiology and Medicine. Furthermore, a new T-cell called the CAR T-cell was developed. The new type of receptor of the CAR T-cell is known as a chimeric antigen receptor (CAR) and functions as a cancer-specific antibody targeting antigens on the cancer cell surface. The CAR T-cell therapy, which is intended to completely cure cancer, received FDA approval. The CAR T-cell was demonstrated to be effective in achieving clinical remission of ALL (acute lymphocytic leukemia) patients. The CAR T-cell engineering will advance for applying these therapeutic approaches more broadly to solid cancers such as breast, lung, and head and neck cancers.

In 1993, the SOD1 gene was the first to be associated with ALS. A recent metanalysis found that pathogenic variants in SOD1 account for approximately 15–30% of fALS (see Chapter 9) and fewer than 2% of sALS. In 2008, dominant mutations in the TARDBP gene were identified as a primary cause of ALS, providing evidence that aberrant TDP-43 (TARDNA-binding protein 43 encoded by the TARDBP gene) could be causative of neurodegeneration.

With the advent of iPS (induced pluripotent stem) cells and transdifferentiated cells, these cells can serve for investigation of disease mechanisms and in the search for new drugs for incurable diseases.

As a result, it is found that the human body protects against disability of health by its immune system, and human healthcare has been held by various biomedical treatments developed by great medical scientists.

This book describes features of cells and the cell cycle, as an introductory explanation. Then, it describes the immune system's feature of protecting against microorganisms causing infectious diseases, and vaccination preventing future infection. And then, it describes the various biomedical treatments for diseases, especially fatal diseases such as cancers and ALS, developed on the basis of molecular biology and immunobiology.

As the author, I am eager that readers with extreme enthusiasm for medicine will intend to develop new therapies, to reply to patients' expectation for the arrival of the days when suffering is dispelled by new therapies.

Explanations are given for medical terminologies. For example, section E1.2 represents the second explanation in Chapter 1. Citation [1]:1015 means that page 1015 of the reference [1] is referred to.

Acknowledgments

The author would like to express his appreciation to Doctor Ms. Nozomi Tani at the Kyoto Kuramaguchi Medical Center and associate professor Mr. Tomonori Yaguchi, M.D., Ph.D. at the Center for Cancer Immunotherapy and Immunobiology (CCII) in Kyoto University, for their instructive comments on his manuscript.
The author appreciates Project Editor Ms. Christie Naglieri, Senior Acquisition Editor Mr. C.R. Crumly, Ph.D., Editorial Assistant Ms. Austin Decker, Mr. Christian Munoz, Lead Project Management Mr. Balaji Karuppanan, Mr. Todd Perry, and all the members in the publisher, for facilitating the publication of this book. He also appreciates other authors for the books and the articles referred to in this book.

Tadayoshi Shioyama
In Kyoto September 2026

1 Cells and Genomes

The surface of our planet is populated by organisms which are regarded as curious organized chemical factories that take in nutrients from their surroundings and use these raw materials to generate descendants of themselves. These living organisms appear extraordinarily diverse. Our ancestors, knowing nothing of cells or DNA (deoxyribonucleic acid), saw that all these organisms had something in common. They called that something "life," marveled at it, struggled to define it, and despaired of explaining what it was or how it worked.

The discoveries of the past century have removed the central mystery regarding the nature of life. We are now able to understand that all organisms are made of cells: small, membrane-enclosed units filled with a concentrated aqueous solution of chemicals and endowed with the extraordinary ability to create copies of themselves by growing and then dividing into two.

Because cells are the fundamental units of life, it is cell biology, which is the study of the structures, function, and behavior of cells, that we must look to for answers to questions of what life is and how it works. A deeper understanding of cells and their evolution leads to the ability to tackle the grand historical problems of life on Earth: its mysterious origins, its stunning diversity, and its invasion of every conceivable habitat.

Despite their apparent diversity, organisms are fundamentally similar inside. The whole of biology is a counterpoint between two themes: astonishing variety in individual particulars and astonishing constancy in fundamental mechanisms. In this chapter, we begin by outlining the universal features common to all life on our planet. We then survey the diversity of cells. And we see how it is possible to read, measure, and decipher these specifications to help us accomplish a coherent understanding of all the forms of life from the smallest to the greatest.

The human body is a multicellular organism and is formed by the aggregation of more than 10^{13} (ten trillion) cells. The cells of the human body are eukaryotic cells; that is, they have a nucleus (E1.5). There is a chromosome whose core is formed by DNA in the nucleus. In a cytosol outside the nuclei of eukaryotic cells, there exist mitochondria (E1.6) and so on [1].

1.1 UNIVERSAL FEATURES OF CELLS

1.1.1 The Universal Features of Cells on Earth

It is estimated that more than 10 million – perhaps 100 million – organisms populate the earth today. Each species is different, and each reproduces itself faithfully, yielding progeny that belong to the same species. The parent organism hands down information specifying, in extraordinary detail, the characteristics with which the offspring shall be endowed. This phenomenon of heredity is central to the definition

DOI: 10.1201/9781003780663-1

of life: it distinguishes life from other processes, such as the growth of a crystal, or the formation of waves on water, in which there is nothing similar to organisms with the same type of link between the peculiarities of parents and the peculiarities of offspring. The living organism must consume free energy to create and maintain its organization, employing the free energy to drive a hugely complex system of chemical processes that are specified by hereditary information.

Most living organisms are single cells. Others, such as human beings, are vast multicellular organisms in which groups of cells perform specialized functions linked by intricate systems of communication. However, even for the aggregate of more than 1013 cells that form a human body, the whole organism has been generated from a single cell by cell divisions. Thus, the single cell is the vehicle for all of the hereditary information that defines each species. This includes the machinery to gather raw materials from the environment and construct from them a new cell in its own image, complete with a new copy of its hereditary information.

1.1.2 All Cells Store Their Hereditary Information in DNA

Through computers, we have been made familiar with the concept of information as a measurable quantity. Living cells, like computers, store information, and it is estimated that they have been evolving and diversifying for over 3.5 billion years. All living cells on the earth store their hereditary information in the form of double-stranded molecules of DNA – long, unbranched, paired polymer chains, formed always of the same four types of monomers. These monomers, known as nucleotides, have nicknames drawn from a four-letter alphabet (A, T, C, G) and they are strung together in a long linear sequence that encodes the genetic information, just as the sequence of 1s and 0s encodes the information in a computer file. A normal DNA molecule consists of two complementary strands. The nucleotides within each strand are linked by strong covalent chemical bonds; the complementary nucleotides on opposite strands are held together more weakly by hydrogen bonds.

The strategy for deciphering all of the hereditary information is as follows. A DNA fragment is inserted into a bacterial plasmid with the enzyme DNA ligase. The entire collection of DNA fragments is ligated into plasmid vectors. These recombinant plasmids are then introduced into *E. coli* at a concentration that ensures that no more than one plasmid molecule is taken up by each bacterium. The collection of cloned plasmid molecules is known as a DNA library. Because the DNA fragments were derived directly from the chromosomal DNA of the organism of interest, the resulting collection, called a genomic library, will represent the entire genome of the organism. A plasmid is a separate intracellular fragment of bacterial DNA. Thus, the information will be successfully read, interpreted, and copied [1]:468–469. Using chemical methods, scientists have learned how to read out the complete sequence of monomers in any DNA molecule – extending for many millions of nucleotides – and thereby decipher all of the hereditary information that each organism contains [1]:3.

As shown in Figure 1.1, DNA (deoxyribonucleic acid) is made from simple subunits, called nucleotides, each consisting of a sugar phosphate molecule with a nitrogen-containing side group, or base, attached to it. The bases are of four types (adenine, guanine, cytosine, and thymine), corresponding to four distinct nucleotides, labeled

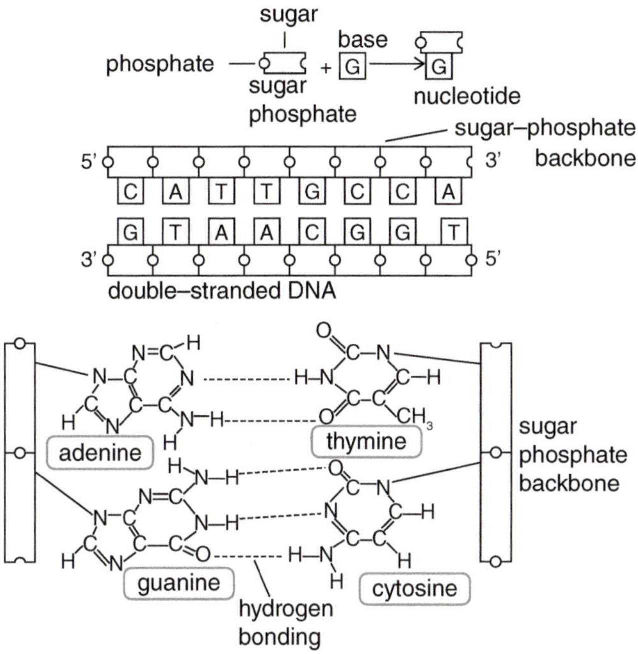

FIGURE 1.1 Base pairs in DNA.

A, G, C, and T. A single strand of DNA consists of nucleotides joined together by sugar phosphate linkages. A normal DNA molecule consists of two such complementary strands, each of which is formed by strongly bound nucleotides; the complementary nucleotides on opposite strands are held together more weakly, by hydrogen bonds. The two strands twist around each other to form a double helix.

1.1.3 DNA Replication

The mechanisms that make life possible depend on the structure of the double-stranded DNA molecule. Each monomer in a single DNA strand – that is, each nucleotide – consists of two parts: a sugar (deoxyribose) with a phosphate group attached to it, and a base, which may be either adenine (A), guanine (G), cytosine (C), or thymine (T) (Figure 1.1). Each sugar is linked to the next via the phosphate group, creating a polymer chain composed of a repetitive sugar-phosphate backbone with a series of bases protruding from it. The DNA polymer is extended by adding monomers at one end. For a single isolated strand, these monomers can, in principle, be added in any order. In the living cell, however, DNA is not synthesized as a free strand in isolation, but on a template formed by a preexisting DNA strand (Figure 1.2). The bases protruding from the existing strand bind to bases of the strand being synthesized, according to a strict rule defined by the complementary structures of the bases: A binds to T, and C binds to G. Thus, base-pairing holds fresh monomers in place and thereby controls

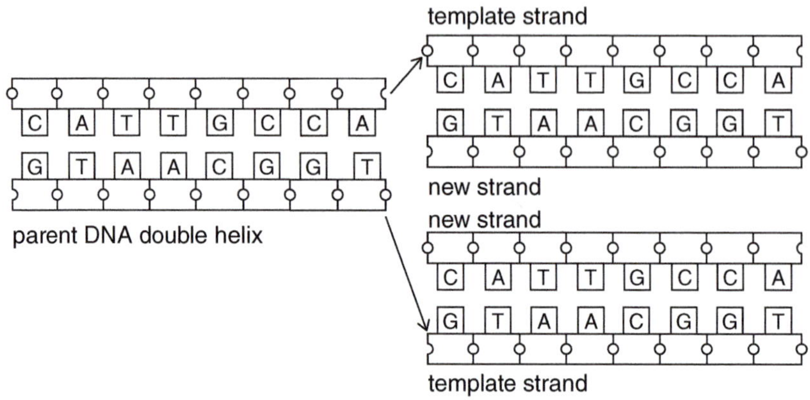

FIGURE 1.2 DNA replication.

the selection of which one of the four monomers shall be added to the growing strand next. In this way, a double-stranded structure is created, consisting of two exactly complementary sequences of As, Cs, Ts, and Gs. The two strands twist around each other, forming a DNA double helix.

The bonds between the base pairs are weak compared with the sugar-phosphate links, allowing the two DNA strands to be pulled apart without breakage of their backbones. Each strand can serve as a template, in the way just described, for the synthesis of a fresh DNA strand complementary to itself – a fresh copy, that is, of the hereditary information (Figure 1.2).

In different types of cells, this process of DNA replication occurs at different rates, with different controls to start it or stop it, and different auxiliary molecules to help it along. But the basics are universal: DNA is the information store for heredity, and templated polymerization is the way in which this information is copied throughout the living world.

1.2 TRANSCRIBING DNA AND TRANSLATING

1.2.1 All Cells Transcribe Portions of DNA into the Same Intermediary Form: RNA

To carry out its information-bearing function, DNA must do more than copy itself; that is, it must also let the information guide the synthesis of other molecules in the cell, leading first and foremost to the production of two other key classes of polymers: RNAs and proteins. The mechanism of synthesizing the polymers – RNAs (ribonucleic acids) and proteins – is the same in all living organisms. The process begins with a templated polymerization called transcription, in which segments of the DNA sequence are used as a template for the synthesis of shorter molecules of the closely related polymer ribonucleic acid or RNA. Later, in the more complex process of translation, many of these RNA molecules direct the synthesis of polymers of a radically different chemical class – the proteins.

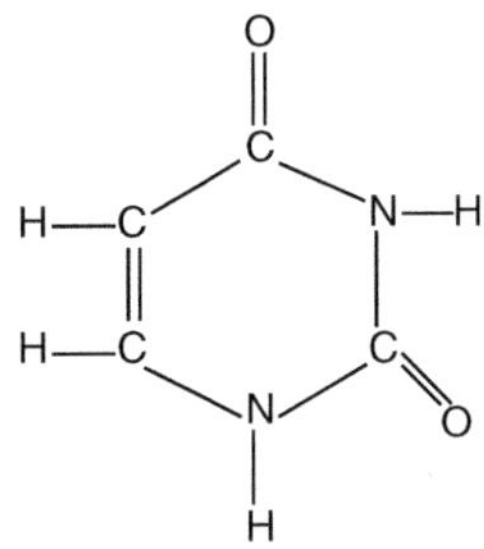

FIGURE 1.3 Uracil.

In RNA, the backbone is formed of a slightly different sugar from that of DNA – ribose instead deoxyribose – and one of the four bases is slightly different – uracil (U) in place of thymine (T) (Figure 1.3). But the other three bases – A, C, and G – are the same, and all four bases pair with their complementary counterparts in DNA – the A, U, C, and G of RNA with the T, A, G, and C of DNA. During transcription, the RNA monomers are lined up and selected for polymerization on a template strand of DNA, just as DNA monomers are selected during replication. The outcome is a polymer molecule whose sequence of nucleotides faithfully represents a portion of the cell's genetic information, even though it is written in a slightly different alphabet – consisting of RNA monomers instead of DNA monomers. Thus, whereas the cell's archive of genetic information in the form of DNA is fixed and sacrosanct, these RNA transcripts are mass-produced and disposable. These transcripts function as intermediates in the transfer of genetic information. Most notably, they serve as messenger RNA (mRNA) molecules that guide the synthesis of proteins according to the genetic instructions stored in the DNA.

RNA molecules have distinctive structures that can also give them other specialized chemical capabilities. Being single-stranded, their backbone is flexible, so that the polymer chain can bend back on itself to allow one part of the molecule to form weak bonds with another part of the same molecule. This occurs when segments of the sequence are locally complementary: a . . .GGGG . . . segment, for example, will tend to associate with a . . .CCCC . . . segment. These types of internal associations can cause an RNA chain to fold up into a specific shape that is dictated by its sequence. As shown in Figure 1.4, nucleotide pairing between different regions of the same RNA polymer chain causes the molecule to adopt a distinctive shape. The shape of the RNA molecule, in turn, may enable it to recognize other molecules by binding to them selectively – and even, in certain cases, to catalyze chemical changes in the molecules that are bound. In fact, some chemical reactions catalyzed by RNA molecules are crucial for several of the most ancient and fundamental processes in living cells, and it has been suggested that an extensive catalysis by RNA played a central part in the early evolution of life. That is, because an RNA molecule is single-stranded, an RNA chain can fold up into a specific shape. The shape of the RNA molecule may enable it to recognize other molecules and to catalyze chemical changes in the molecules bound.

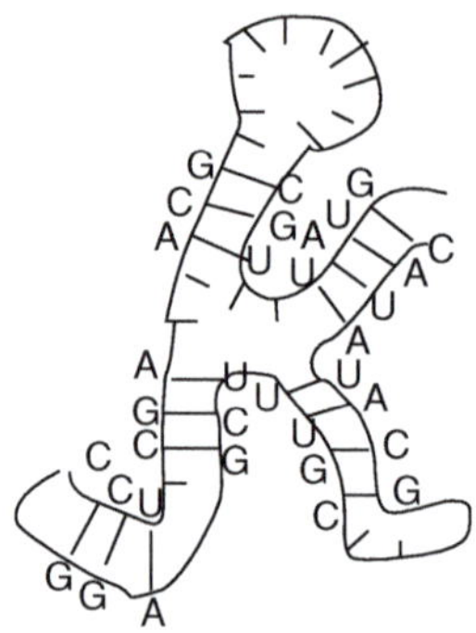

FIGURE 1.4 The conformation of an RNA molecule; RNA can fold into specific structures [1]:5.

1.2.2 mRNA and Noncoding RNAs

The majority of genes carried in a cell's DNA specify the amino acid sequence of proteins; the RNA molecules that are copied from these genes (which ultimately direct the synthesis of proteins) are called messenger RNA (mRNA) molecules. The final product from other genes than those specifying amino acids, however, is the RNA molecule itself.

These RNAs are known as noncoding RNAs because they do not code for protein. In the well-studied single-celled eukaryote, the yeast *Saccharomyces cerevisiae*, over 1,200 genes (more than 15% of the total) produce RNA as their final product. Humans may produce on the order of ten thousand noncoding RNAs. These RNAs, like proteins, serve as enzymatic, structural, and regulatory components for a wide variety of processes in the cell. Although many of the noncoding RNAs are still mysterious, it is known that small nuclear RNA (snRNA) molecules direct the splicing of pre-mRNA to form mRNA; it is known that ribosomal RNA (rRNA) molecules form the core of ribosomes; it is known that transfer RNA (tRNA) molecules form the adaptors that select amino acids and hold them in place on a ribosome for incorporation into protein; and it is known that micro RNA(miRNA) molecules regulate gene expression by blocking translation of specific mRNAs and cause their degradation. That is, among noncoding RNAs, there are snRNA directing the splicing of pre-mRNA, rRNA forming the core of ribosomes, tRNA forming the adaptors, which select amino acids and hold them in place on a ribosome, and miRNA regulating gene expression.

1.2.3 All Cells Use Proteins as Catalysts

Protein molecules are long unbranched polymer chains, formed by stringing together monomeric building blocks drawn from a standard repertoire that is the same for all living cells. Proteins carry information in the form of a linear sequence of symbols, in the same way as a human message is written in an alphabetic script. There are many different protein molecules in each cell, and – leaving out the water – they form most of the cell's mass.

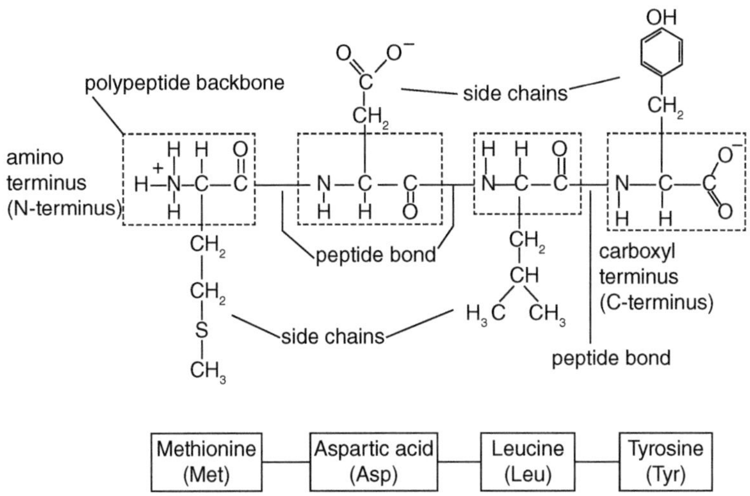

FIGURE 1.5 The components of a protein. A protein consists of a polypeptide backbone with attached side chains. Each type of protein differs in its sequence and number of amino acids; therefore, it is the sequence of the chemically different side chains that makes each protein distinct [1]:110.

The monomers of protein, the amino acids, are quite different from those of DNA and RNA, and there are 20 types instead of 4; that is, A, T, C, and G or A, U, C, and G. Each amino acid is built around the same core structure (polypeptide backbone) through which it can be linked in a standard way to any other amino acid in the set; attached to this core is a side group (side chains) that gives each amino acid a distinctive chemical character (Figure 1.5).

Each of the protein molecules is a polypeptide, created by joining its amino acids in a particular sequence. Through billions of years of evolution this sequence has been selected to give the protein a useful function. Thus, by folding into a precise three-dimensional form with reactive sites on its surface, these amino-acid polymers can bind with high specificity to other molecules and can act as enzymes to catalyze reactions that make or break covalent bonds. In this way, they direct the vast majority of chemical processes in the cell. That is, the protein is a catalyst acting as enzymes in the cell.

Proteins have many other functions as well – maintaining structures, generating movements, sensing signals, and so on – each protein molecule performing a specific function according to its own genetically specified sequence of amino acids. Proteins, above all, are the main molecules that put the cell's genetic information into action.

Thus, polynucleotides specify the amino acid sequences of proteins. Proteins, in turn, catalyze many chemical reactions, including those DNA polymerases by which new DNA molecules are synthesized. From the most fundamental point of view, a living cell is a self-replicating collection of catalysts that takes in food, processes this food to derive both the building blocks and energy needed to make more catalysts, and discards the materials left over as waste. A feedback loop that connects proteins catalyzing DNA synthesis and polynucleotides concerning DNA molecules forms the basis for this autocatalytic, self-reproducing behavior of living organisms.

1.2.4 Translating RNA into Protein

In this subsection, we describe that all cells translate RNA into protein in the same way. In the 1950s when the double-stranded structure of DNA was first revealed as the basis of heredity, it was a complete mystery how the information in DNA specifies the production of proteins. But in the intervening years, scientists have discovered the elegant mechanisms involved. Translation of genetic information from the 4-letter alphabet of polynucleotides into the 20-letter alphabet of proteins is a complex process.

It turns out that the information in the sequence of a messenger RNA molecule is read out in groups of three nucleotides at a time: each triplet of nucleotides, or codon, specifies (codes for) a single amino acid in a corresponding protein. Since the number of distinct triplets that can be formed from four nucleotides is 43, there are 64 possible codons, all of which occur in nature. However, there are only 20 naturally occurring amino acids, meaning that there are necessarily many cases in which several codons correspond to the same amino acid. This genetic code is read out by a special class of small RNA molecules, the transfer RNAs (tRNAs). Each type of tRNA becomes attached at one end to a specific amino acid, and displays at its other end a specific sequence of three nucleotides – an anticodon – that enables it to recognize, through base pairing, a particular codon or subset of codons in mRNA.

The intricate chemistry that enables these tRNA to translate a specific sequence of A, C, G, and U nucleotides in an mRNA molecule into a specific sequence of amino acids in a protein molecule occurs on the ribosome, a large multimolecular machine composed of both protein and ribosomal RNA (rRNA) [1]:7.

The flow of genetic information in cells is from DNA to mRNA to protein. All cells, from bacteria to humans, express their genetic information in this way – a principle so fundamental that it is termed the central dogma of molecular biology.

1.3 TRANSCRIPTION AND RNA PROCESSING

1.3.1 RNA Polymerase Carrying Out Transcription

The enzymes that perform transcription are called RNA polymerases. Like the DNA polymerase that catalyzes DNA replication, RNA polymerases catalyze the formation of the phosphodiester bonds that link the nucleotides together to form a linear chain. The RNA polymerase moves stepwise along the DNA, unwinding the DNA helix just ahead of the active site for polymerization to expose a new region of the template strand for complementary base-pairing. In this way, the growing RNA chain is extended by one nucleotide at a time in the 5'-to-3' direction (Figure 1.6). The substrates are ribonucleoside triphosphates (ATP, CTP, UTP, and GTP); as in DNA replication, the hydrolysis of high-energy bonds provides the energy needed to drive the reaction forward.

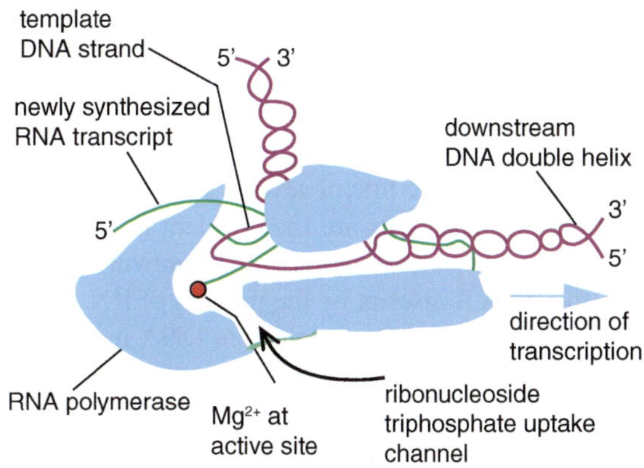

FIGURE 1.6 DNA is transcribed by the enzyme RNA polymerase. The RNA polymerase (pale blue) moves stepwise along the DNA, unwinding the DNA helix at its active site indicated by the Mg^{2+} (red), which is required for catalysis. As it progresses, the polymerase adds nucleotides, one by one, to the RNA chain at the polymerization site, using an exposed DNA strand as a template. The RNA transcript is thus a complementary copy of one of the two DNA strands. A short region of DNA/RNA helix (approximately nine nucleotide pairs in length) is formed only transiently, and a "window" of DNA/RNA helix moves along the DNA with the polymerase as the DNA double helix reforms behind it. The incoming nucleotides are in the form of ribonucleoside triphosphates (ATP, UTP, CTP, and GTP), and energy stored in their phosphate–phosphate bonds provides the driving force for the polymerization reaction [1]:304.

The almost immediate release of the RNA strand from the DNA as it is synthesized means that many RNA copies can be made from the same gene in a relatively short time, with the synthesis of additional RNA molecules being started before the previous RNA molecules are completed. When RNA polymerase molecules follow hard on each other's heels in this way, each moving at about 50 nucleotides per second, over a thousand transcripts can be synthesized in an hour from a single gene. Although RNA polymerase catalyzes essentially the same chemical reaction as DNA polymerase, there are some important differences between the activities of the two enzymes. Firstly, and most obviously, RNA polymerase catalyzes the linkage of ribonucleotides, not deoxyribonucleotides. Secondly, unlike the DNA polymerases involved in DNA replication, RNA polymerases can start an RNA chain without a primer. This difference is thought possible because transcription need not be as accurate as DNA replication. RNA polymerases make about one mistake for every 104 nucleotides copied into RNA (compared with an error rate for direct copying by DNA polymerase of about one in 107 nucleotides); and the consequences of an error in RNA transcription are much less significant because RNA does not permanently store genetic information in cells. That is, because RNA does not permanently store genetic information in cells, transcription need not be as accurate as DNA replication.

1.3.2 Transcription Elongation in Eukaryotes Is Tightly Coupled to RNA Processing

Bacterial mRNAs are synthesized by the RNA polymerase starting and stopping at specific spots on the genome. The situation in eukaryotes is substantially different. In particular, transcription is only the first of several steps needed to produce a mature mRNA molecule. Other critical steps are the covalent modification of the primary RNA transcribed by RNA processing, such as the removal of intron sequences that are discarded from the RNA transcript by the process of RNA splicing (Figure 1.7). That is, the primary RNA after transcription from DNA includes exons and introns (noncoding sequences). So the primary RNA is transcribed by RNA processing in order to remove introns.

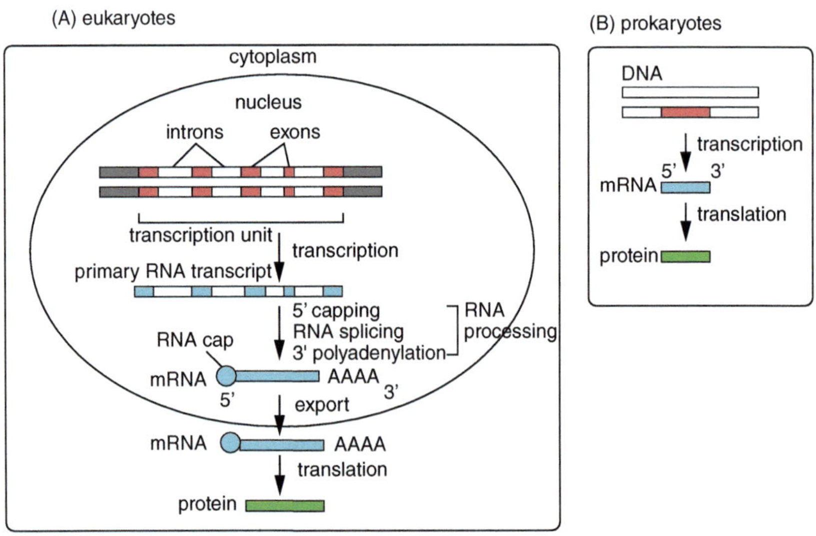

FIGURE 1.7 Comparison of the steps leading from gene to protein in eukaryotes and prokaryotes (bacteria). The final level of a protein in the cell depends on the efficiency of each step and on the rates of degradation of the RNA and protein molecules. (A) In eukaryotic cells, the mRNA molecule resulting from transcription contains both coding (exon) and noncoding (intron) sequences. Before it can be translated into protein, the two ends of the RNA are modified, the introns are removed by an enzymatically catalyzed RNA splicing reaction, and the resulting mRNA is transported from the nucleus to the cytoplasm. For convenience, the steps in this figure are depicted as occurring one at a time; in reality, many occur concurrently. For example, the RNA cap is added and splicing begins before transcription has been completed. Because of the coupling between transcription and RNA processing, intact primary transcripts become 5' capped mRNA with poly-A tail after RNA processing. (B) In prokaryotes, the production of mRNA is much simpler. The 5' end of an mRNA molecule is produced by the initiation of transcription, and the 3' end is produced by the termination of transcription. Since prokaryotic cells lack a nucleus, transcription and translation take place in a common compartment, and the translation of the bacterial mRNA often begins before its synthesis has been completed [1]:315.

Both ends of eukaryotic mRNAs are modified: by capping on the 5' end and by polyadenylation of the 3' end. These special ends allow the cell to assess whether both ends of the mRNA molecule are present (and if the message is therefore intact) before it exports the RNA from the nucleus and translates it into protein. RNA splicing joins together the different portions of a protein-coding sequence, and it provides eukaryotes with the ability to synthesize several different proteins from the same genes due to removing introns in different ways.

A simple strategy has evolved to couple all of the foregoing RNA processing steps to transcription elongation. A key step in transcription initiation by RNA polymerase II is the phosphorylation of RNA polymerase II tail, called the CTD (C-terminal domain). This phosphorylation, which proceeds gradually as the RNA polymerase initiates transcription and moves along the DNA, not only helps dissociate the RNA polymerase II from other proteins present at the start point of transcription, but also allows a new set of proteins to associate with the RNA polymerase tail that function in transcription elongation and RNA processing. Some of these processing proteins are thought to "hop" from the polymerase tail onto the nascent RNA molecule to begin processing it as it emerges from the RNA polymerase. Thus, we can view RNA polymerase II in its elongation mode as an RNA factory that not only moves along the DNA synthesizing an RNA molecule, but also processes the RNA that it produces [1]:316. That is, RNA polymerase II not only initiates RNA transcription, but also begins RNA processing.

1.3.3 RNA Capping Being the First Modification of Eukaryotic Pre-mRNAs

As soon as RNA polymerase II has produced about 25 nucleotides of RNA, the 5' end of the new RNA molecule is modified by addition of a cap that consists of a modified guanine nucleotide (Figure 1.8B). Three enzymes – a phosphatase, a guanyl transferase, and a methyl transferase – perform the capping reaction.

The 5'-methyl cap signifies the 5' end of eukaryotic mRNAs, and this landmark helps the cell to distinguish mRNAs from the other types of RNA molecules present in the cell. For example, RNA polymerases I and III produce uncapped RNAs during transcription, in part because these polymerases lack a CTD. In the nucleus, the cap binds a protein complex called CBC (cap-binding complex), which helps a future mRNA be further processed and exported. The 5'-methyl cap also has an important role in the translation of mRNAs in the cytosol.

1.3.4 RNA Splicing Process Removing Intron Sequences from Newly Transcribed pre-mRNAs

The protein-coding sequences of eukaryotic genes are typically interrupted by non-coding intervening sequences (introns). This feature of eukaryotic genes, discovered in 1977, came as a surprise to scientists, who until that time had been familiar only with bacterial genes, which typically consist of a continuous stretch of coding DNA that is directly transcribed into mRNA. In marked contrast, eukaryotic genes were

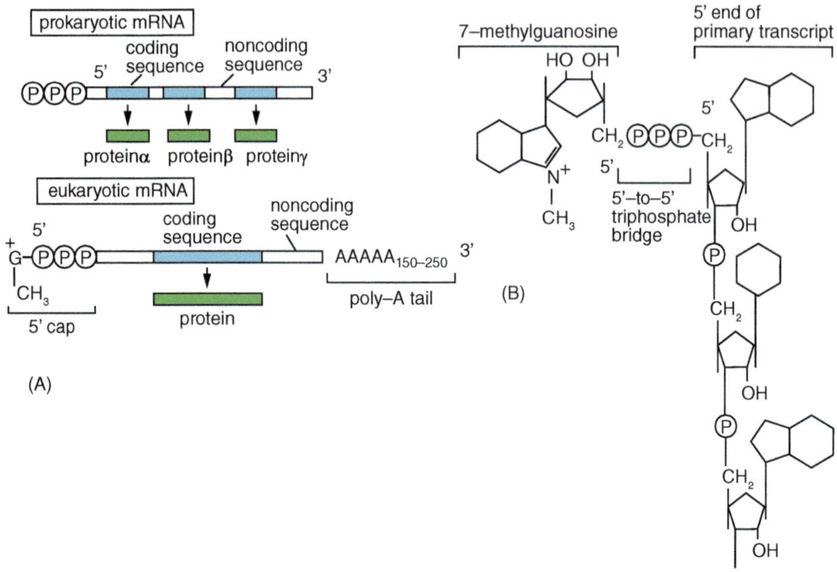

FIGURE 1.8 A comparison of the structures of prokaryotic and eukaryotic mRNA molecules. (A) The 5' and 3' ends of a bacterial mRNA are the unmodified ends of the chain synthesized by the RNA polymerase, which initiates and terminates transcription at those points, respectively. The corresponding ends of a eukaryotic mRNA are formed by adding a 5' cap, and by cleavage of the pre-mRNA transcript near the 3' end, and the addition of a poly-A tail, respectively. The figure also illustrates another difference between the prokaryotic and eukaryotic mRNAs: bacterial mRNAs can contain the instructions for several different proteins, whereas eukaryotic mRNAs nearly always contain the information for only a single protein. (B) The structure of the cap at the 5' end of eukaryotic mRNA molecules. Note the unusual 5'-to-5' linkage of the 7-methyl G to the remainder of the RNA. Many eukaryotic mRNAs carry an additional modification: methylation of the 2'-hydroxyl group of the ribose sugar at the 5' end of the primary transcript ([1], Figure 6.23:316–317).

found to be broken up into small pieces of coding sequences (exons) interspersed with much longer intervening sequences or introns; thus, the coding portion of a eukaryotic gene is often only a small fraction of the length of the gene.

Both intron and exon sequences are transcribed into RNA. The intron sequences are removed from the newly synthesized RNA through the process of RNA splicing. The vast majority of RNA splicing that takes place in cells, functions in the production of mRNA, and our discussion of splicing focuses on this so-called precursor-mRNA (or pre-mRNA) splicing.

Each splicing event removes one intron, proceeding through two sequential phosphoryl-transfer reactions known as transesterifications ([1]:318); these join two exons together while removing the intron between them as a "lariat" (Figure 1.9). The machinery that catalyzes pre-mRNA splicing is complex, consisting of five

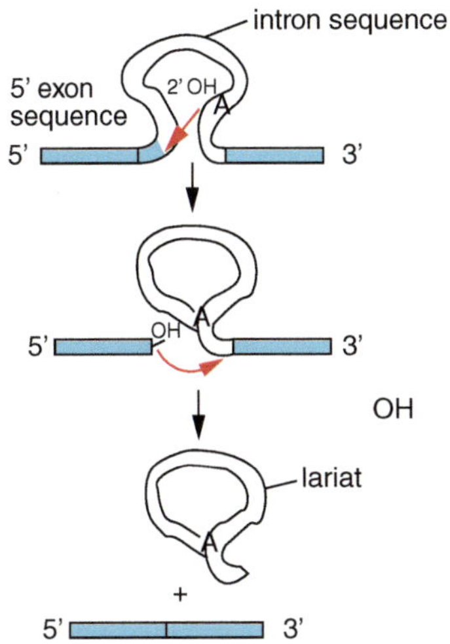

FIGURE 1.9 The pre-mRNA splicing reaction. In the first step, a specific adenine nucleotide in the intron sequence (indicated by A) attacks the 5' end of the intron and cuts the sugar-phosphate backbone of the RNA at this point. The cut 5' end of the intron becomes covalently linked to the adenine nucleotide, thereby creating a loop in the RNA molecule. The released free 3'-OH end of the exon sequence then reacts with the start of the next exon sequence, joining the two exons together and releasing the intron sequence in the shape of a lariat. The two exon sequences thereby become joined into a continuous coding sequence. The released intron sequence is eventually broken down into single nucleotides, which are recycled [1]:318.

additional RNA molecules and several hundred proteins, and it hydrolyzes many ATP molecules per splicing event. This complexity ensures that splicing is accurate, while at the same time being flexible enough to deal with the enormous variety of introns found in a typical eukaryotic cell.

It may seem wasteful to remove large numbers of introns by RNA splicing. In attempting to explain why it occurs, scientists have pointed out that the exon-intron arrangement would seem to facilitate the emergence of new and useful proteins over an evolutionary time scale. Thus, the presence of numerous introns in DNA allows genetic recombination to readily combine the exons of different genes, enabling genes for new proteins to evolve more easily by the combination of parts of preexisting genes. The observation that many proteins in present-day cells resemble patchworks composed from a common set of protein domains ([1]:121–122) supports this idea.

RNA splicing also has a present-day advantage. The transcripts of many eukaryotic genes (estimated at 95% of genes in humans) are spliced in more than one way, thereby allowing the same gene to produce a corresponding set of different proteins, because the primary transcript can be spliced in different ways.

1.4 TRANSLATION FROM MRNA (MESSENGER RNA)

1.4.1 From RNA to Protein

Most genes in a cell produce mRNA molecules that serve as intermediaries on the pathway to proteins. In this section, we examine how the cell converts the information carried in an mRNA molecule into a protein molecule. This feat of translation was a strong focus of attention for biologists in the late 1950s, when it was posed as the coding problem: how is the information in a linear sequence of nucleotides in RNA translated into the linear sequence of a chemically quite different set of units – the amino acids in proteins? Here was a cryptogram set up by nature that, after more than 3 billion years of evolution, could finally be solved by one of the products of evolution – human beings. And indeed, in the year 2000, the structure of the elaborate machinery by which cells read this code – the ribosome – was finally revealed in atomic detail.

1.4.2 Decoding an mRNA Sequence in Sets of Three Nucleotides

The information present in an mRNA produced by transcription and RNA processing is used to synthesize a protein. Transcription is simple to understand as a means of information transfer: since DNA and RNA are chemically and structurally similar, the DNA can act as a direct template for the synthesis of RNA by complementary base-pairing. As the term transcription signifies, it seems as if a message written out by hand is being converted into a typewritten text. The language itself and the form of the message do not change, and the symbols used are closely related.

By contrast, the conversion of the information in RNA into protein represents a translation of the information into another language that uses quite different symbols. Moreover, since there are only 4 different nucleotides in mRNA and 20 different types of amino acids in a protein, this translation cannot be explained by a direct one-to-one correspondence between a nucleotide in RNA and an amino acid in protein. The nucleotide sequence of a gene, through the intermediary of mRNA, is translated into the amino acid sequence of a protein by rules that are known as the genetic code. This code was deciphered in the early 1960s. The sequence of nucleotides in the mRNA molecule is read in consecutive groups of three nucleotides. RNA is a linear polymer of four different nucleotides, so there are 43=64 possible combinations of three nucleotides: the triplet AAA, AUA, AUG, and so on. However, only 20 different amino acids are commonly found in proteins. Either some nucleotide triplets are never used, or the code is redundant and some amino acids are specified by more than one triplet. The second possibility is, in fact, the correct one, as shown by the completely deciphered genetic code in Figure 1.10. Each group of three consecutive nucleotides in RNA is called a codon and each codon specifies either one amino acid or a stop to the translation process.

In principle, an RNA sequence can be translated in any one of three different reading frames, depending on where the decoding process begins (Figure 1.11). However, only one of the three possible reading frames in an mRNA encodes the required protein.

	AGA									UUA					AGC					
	AGG									UUG					AGU					
GCA	CGA						GGA			CUA				CCA	UCA	ACA			GUA	
GCC	CGC						GGC		AUA	CUC				CCC	UCC	ACC			GUC	UAA
GCG	CGG	GAC	AAC	GGC	GAA	CAA	GGG	CAC	AUC	CUG	AAA		UUC	CCG	UCG	ACG		UAC	GUG	UAG
GCU	CGU	GAU	AAU	UGU	GAG	CAG	GGU	CAU	AUU	CUU	AAG	AUG	UUU	CCU	UCU	ACU	UGG	UAU	GUU	UGA
Ala	Arg	Asp	Asn	Cys	Glu	Gln	Gly	His	Ile	Leu	Lys	Met	Phe	Pro	Ser	Thr	Trp	Tyr	Val	Stop
A	R	D	N	C	E	Q	G	H	I	L	K	M	F	P	S	T	W	Y	V	

FIGURE 1.10 Correspondence between amino acid and genetic code expressed by 4 nucleic acids C, G, A, and U [1]:334. The standard abbreviation for each amino acid is presented below its three-letter abbreviation. By convention, codons are always written with the 5'-terminal nucleotide to the left. Most amino acids are represented by more than one codon, and there are some regularities in the set of codons that specify each amino acid: codons for the same amino acid tend to contain the same nucleotides at the first and second positions, and vary at the third position. Three codons do not specify any amino acid but act as termination sites (stop codon), signaling the end of the protein-coding sequence. One codon – AUG – acts both as an initiation codon, signaling the start of a protein-coding message, and also as the codon that specifies methionine.

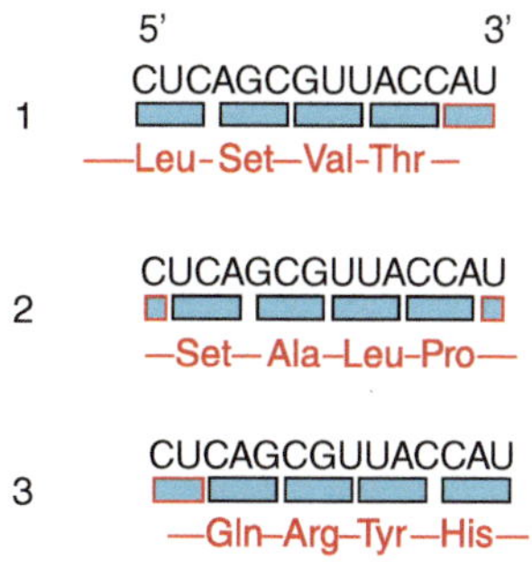

FIGURE 1.11 The three possible reading frames in protein synthesis[1]:334. In the process of translating a nucleotide sequence (blue) into an amino acid sequence (red), the sequence of nucleotides in an mRNA molecule is read from the 5' end to the 3' end in consecutive sets of three nucleotides. In principle, the same RNA sequence can specify three completely different amino acid sequences, depending on the reading frame. In reality, however, only one of these reading frames contains the actual message (see section 1.6).

1.5 tRNA (TRANSFER RNA) MOLECULES AS AN ADAPTOR FOR TRANSLATION

1.5.1 tRNA Molecules Matching Amino Acids to Codons in mRNA

The codons in an mRNA molecule do not recognize the amino acids which are specified by codons: the codon, which is the group of three nucleotides, does not bind directly to the amino acid. Rather, the translation of mRNA into protein depends on adaptor molecules that can recognize and bind both to the codon and at another site on their surface, to the amino acid. These adaptors consist of a set of small RNA molecules known as transfer RNA (tRNA), each about 80 nucleotides in length.

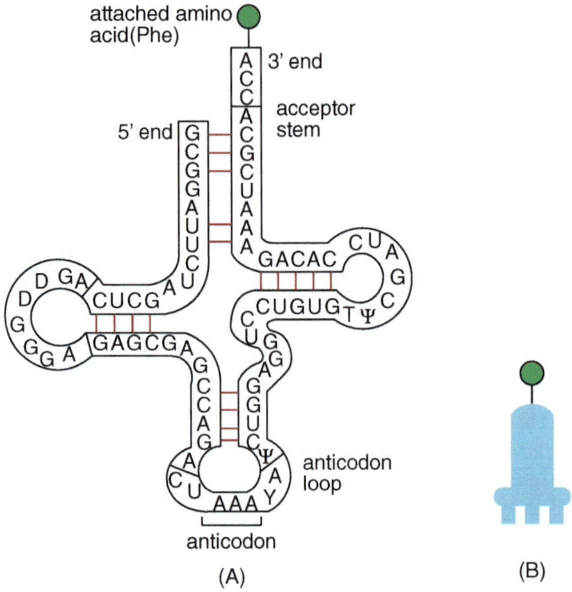

FIGURE 1.12 A tRNA molecule [1]:335. (A) A tRNA specific for the amino acid phenylalanine (Phe) is depicted as a bending structure like a cloverleaf, showing the complementary base-pairing (red lines) that creates the double-helical regions of the molecule. The anticodon is the sequence of three nucleotides that base-pairs with a codon in mRNA. The amino acid matching the codon/anticodon pair is attached at the 3' end of the tRNA. (B) The tRNA icon is shown.

Since RNA molecules are single-stranded molecules, they can fold into precise three-dimensional structures, and the tRNA molecules provide a striking example. Four short segments of the folded tRNA are double-helical, producing a molecule that looks like a cloverleaf when drawn schematically (Figure 1.12A). For example, a 5'-GCUC-3' sequence in one part of a polynucleotide chain can form a relatively strong association with a 5'-GAGC-3' sequence in another region of the same molecule (a part of the left leaf of the tRNA).

tRNA contains some unusual bases, which are produced by chemical modification after the tRNA has been synthesized. For example, the bases denoted by ψ (pseudouridine) and D (dihydrouridine) are derived from uracil.

Two regions of unpaired nucleotides situated at either end of the L-shaped molecule are crucial to the function of tRNA in protein synthesis. One of these regions forms the anticodon, a set of three consecutive nucleotides that pairs with the complementary codon in an mRNA molecule. The other is a short single-stranded region at the 3' end of the molecule; this is the site where the amino acid that matches the codon is attached to the tRNA.

The genetic code is redundant; that is, several different codons can specify a single amino acid. This redundancy implies that for an amino acid some tRNA molecules can base-pair with more than one codon.

Some amino acids have more than one tRNA and some tRNAs are constructed so that they require accurate base-pairing only at the first two positions of the codon and can tolerate a mismatch (or wobble) at the third position. This wobble base-pairing explains why so many of the alternative codons for an amino acid differ only in their third nucleotide.

1.5.2 tRNA Being Modified Before Their Exit from the Nucleus

Like most other eukaryotic RNAs, tRNAs are covalently modified before they are allowed to exit from the nucleus. Eukaryotic tRNAs are synthesized by RNA polymerase III. Both bacterial and eukaryotic tRNAs are typically synthesized as larger precursor tRNAs, which are then trimmed to produce the mature tRNA. In addition, some tRNA precursors (from both bacteria and eukaryotes) contain introns that must be spliced out. This splicing reaction differs chemically from pre-mRNA splicing; rather than generating a lariat intermediate, tRNA splicing uses a cut-and-paste mechanism that is catalyzed by proteins. Trimming and splicing both require the precursor tRNA to be correctly folded into its cloverleaf configuration. Because misfolded tRNA precursors will not be processed properly, the trimming and splicing reactions serve as quality-control steps in the generation of tRNAs.

All tRNAs are modified chemically: nearly 1 in 10 nucleotides in each mature tRNA molecule is an altered version of a standard G, U, C, or A ribonucleotide. Over 50 different types of tRNA modifications are known: a few are shown in Figure 1.13.

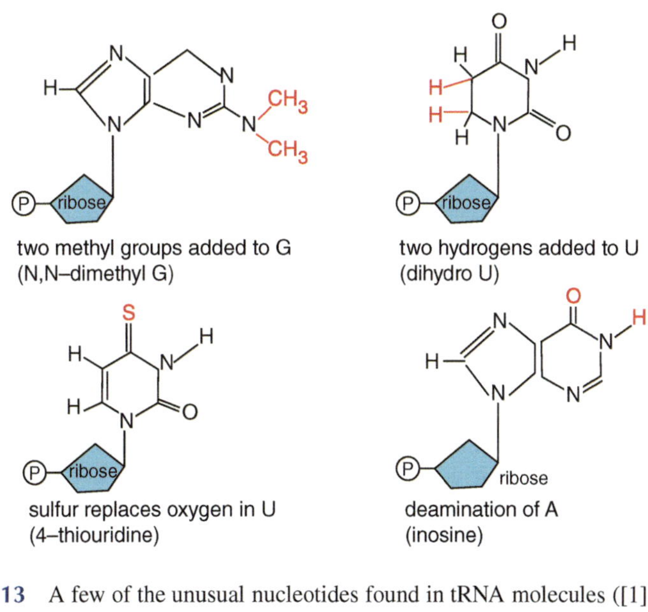

FIGURE 1.13 A few of the unusual nucleotides found in tRNA molecules ([1]:337). These nucleotides are produced by covalent modification of a normal nucleotide after it has been incorporated into an RNA chain. In most tRNA molecules, about 10% of the nucleotides are modified. For example, although the anticodon base for codon U is A, I (inosine) is present at the wobble position in the tRNA anticodon ([1], Figure 6.51:336).

Some of the modified nucleotides – most notably inosine, produced by the deamination of adenosine – affect the conformation and base-pairing of the anticodon and thereby facilitate the recognition of the appropriate mRNA codon by the tRNA molecule. Others affect the accuracy with which the tRNA is attached to the correct amino acid.

1.5.3 Specific Enzymes Coupling Each Amino Acid to Its Appropriate tRNA Molecule

In order to read the genetic code in DNA, cells make a series of different tRNAs. We consider how the tRNA molecule becomes linked to the one amino acid in 20 that is its appropriate partner. Recognition and attachment of the correct amino acid depend on enzymes called aminoacyl-tRNA synthetases (Figure 1.16), which covalently couple each amino acid to its appropriate set of tRNA molecules (Figure 1.14, Figure 1.15). Most cells have a different synthetase enzyme for each amino acid (that is, 20 synthetases in all); one attaches glycine to all tRNAs that recognize codons for glycine, another attaches alanine to all tRNAs that recognize codons for alanine, and so on. Many bacteria, however, have fewer than 20 synthetases, and the

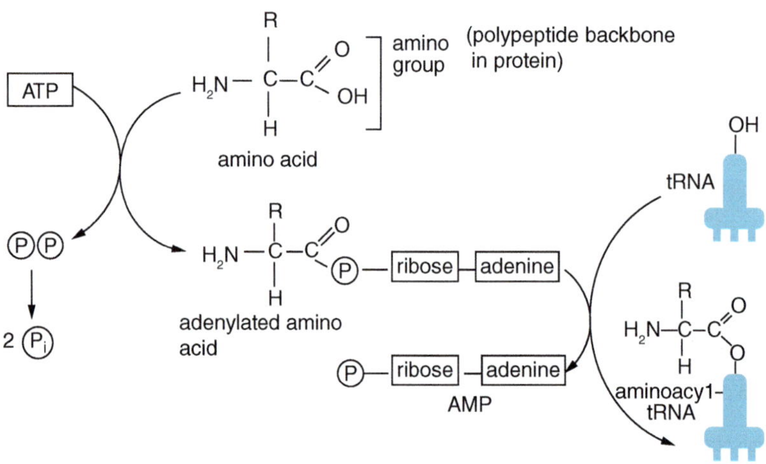

FIGURE 1.14 Amino acid activation by synthetase enzyme [1]:337. An amino acid is activated for protein synthesis by an amino-acyl-tRNA synthetase enzyme in two steps. As indicated, the energy of ATP hydrolysis (Figure 1.16, 1.17) is used to attach each amino acid to its tRNA molecule in a high-energy linkage. The amino acid is first activated through the linkage of its carboxyl group directly to AMP, forming an adenylated amino acid; the linkage of the AMP, normally an unfavorable reaction, is driven by the hydrolysis of the ATP molecule that donates the AMP. Without leaving the synthetase enzyme, the AMP-linked carboxyl group on the amino acid is then transferred to a hydroxyl group on the sugar at the 3' end of the tRNA molecule. This transfer joins the amino acid by an activated ester linkage to the tRNA and forms the final aminoacyl-tRNA molecule. The synthetase enzyme is not shown in this diagram (see Figure 1.16).

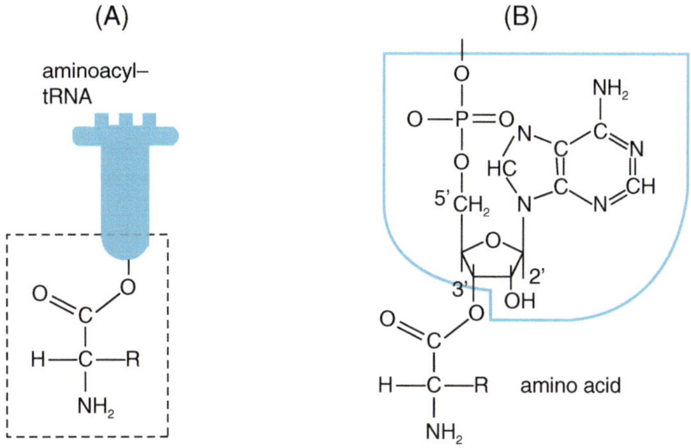

FIGURE 1.15 The structure of the aminoacyl-tRNA [1]:338. The carboxyl end of the amino acid forms an ester bond to ribose. Because the hydrolysis of this ester bond is associated with a large favorable change in free energy, an amino acid held in this way is said to be activated. (A) Schematic drawing of the structure. The amino acid is linked to the nucleotide at the 3' of the tRNA (see Figure 1.12). (B) Actual structure corresponding to the boxed region in (A). There are two major classes of synthetase enzymes: one links the amino acid directly to the 3'OH group of the ribose, and the other links it initially to the 2'-OH group. In the latter case, a subsequent transesterification reaction, in which each splicing event removes one intron, joining two exons together while removing the intron between them as a lariat [1]:318, shifts the amino acid to the 3' position. As in Figure 1.5, the "R group" indicates the side chain of the amino acid.

same synthetase enzyme is responsible for coupling more than one amino acid to the appropriate tRNAs [1]:337.

The synthetase-catalyzed reaction that attaches the amino acid to the 3' end of the tRNA is one of many reactions coupled to the energy-releasing hydrolysis of ATP (see Figure 1.17), and it produces a high-energy bond between the tRNA and the amino acid (Figure 1.16). The energy of this bond is used at a later stage in protein synthesis to link the amino acid covalently to the growing polypeptide chain.

The aminoacyl-tRNA synthetase enzymes and the tRNAs are equally important in the decoding process (Figure 1.16). The first adaptor, the aminoacyl-tRNA synthetase, couples a particular amino acid to its corresponding tRNA. The second adaptor, the tRNA molecule's anticodon, forms base-pairs with appropriate codon on the mRNA. Thus, the genetic code is translated by means of two adaptors.

1.6 PROTEIN SYNTHESIS IN RIBOSOMES

1.6.1 The RNA Message Decoded in Ribosomes

The synthesis of proteins is guided by information carried by mRNA molecules. To maintain the correct reading frame and to ensure accuracy (about 1 mistake every

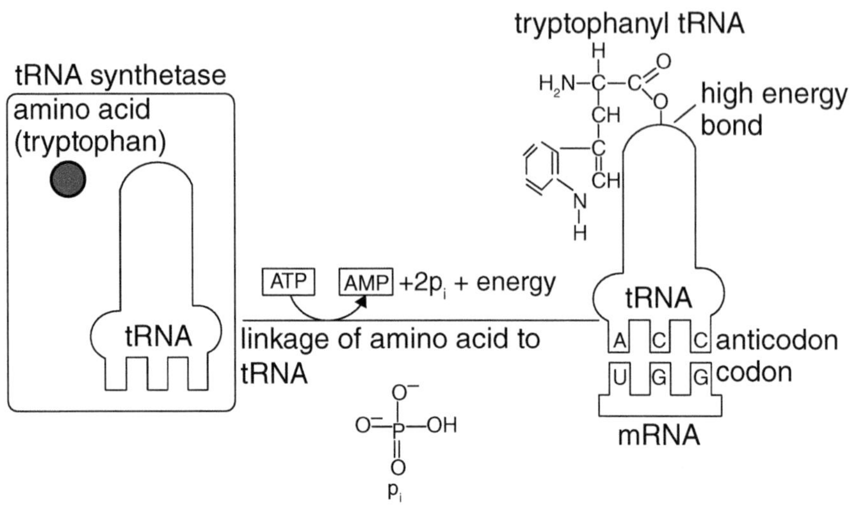

FIGURE 1.16 The genetic code being translated by means of two adaptors that act one after another [1]:338. The first adaptor is the aminoacyl tRNA synthetase, which couples a particular amino acid to its corresponding tRNA: the second adaptor is the tRNA molecule itself, whose anticodon forms base-pairs with the appropriate codon on the mRNA. In the sequence of events shown, the amino acid tryptophan (Trp) is selected by the codon UGG on the mRNA.

10,000 amino acids), protein synthesis is performed in the ribosome, a complex catalytic machine made from more than 50 different proteins (the ribosomal proteins) and several RNA molecules, the ribosomal RNAs (rRNA). A typical eukaryotic cell contains millions of ribosomes in its cytoplasm. The large and small ribosome subunits are assembled at the nucleolus: a ribosome-producing factory which is the site for the processing of rRNAs and their assembly into ribosome subunits [1]:329–331, where newly transcribed and modified rRNA associate with the ribosomal proteins that have been transported into the nucleus after their synthesis in the cytoplasm. These two ribosomal subunits are then exported to the cytoplasm, where they join together to synthesize proteins.

Eukaryotic and bacterial ribosomes have similar structures and functions, being composed of one large and one small subunit that fit together to form a complete ribosome with a mass of several million daltons. The small subunit provides the framework on which the tRNA are accurately matched to the codons of the mRNA, while the large subunit catalyzes the formation of the peptide bonds that link the amino acids together into a polypeptide chain.

When not actively synthesizing proteins, the two subunits of the ribosome are separate. They join together on an mRNA molecule, usually near its 5' end, to initiate the synthesis of a protein. The mRNA is then pulled through the ribosome, three nucleotides at a time. As its codon enters the core of the ribosome, the mRNA nucleotide sequence is translated into an amino acid sequence using the tRNA as an adaptor to add each amino acid in the correct sequence to the growing end of the

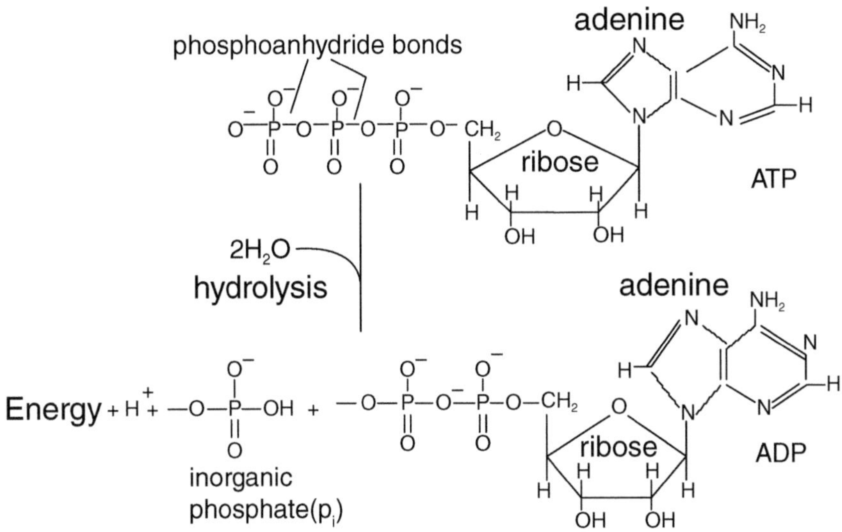

FIGURE 1.17 The hydrolysis of ATP to ADP and inorganic phosphate [1]:65. The two outermost phosphates in ATP are held to the rest of the molecule by high-energy phosphoanhydride bonds and are readily transferred. As indicated, water can be added to ATP to form ADP and inorganic phosphate (p_i). Hydrolysis of the terminal phosphate of ATP yields between 46 and 54 kJ/mole of usable energy, depending on the intracellular conditions. The large negative ΔG of this reaction arises from several factors: release of the terminal phosphate group removes an unfavorable repulsion between adjacent negative charges, and the inorganic phosphate ion (p_i) released is stabilized by resonance and by favorable hydrogen-bond formation with water. ΔG is free energy change in a chemical reaction. ΔG is negative in an energetically favorable reaction. ΔG is positive in an energetically unfavorable reaction [1]:61.

polypeptide chain. When a stop codon is encountered, its ribosome releases the finished protein, and its two subunits separate again. These subunits can then be used to start the synthesis of another protein on another mRNA molecule. Ribosomes operate with remarkable efficiency: in one second, a eukaryotic ribosome adds 2 amino acids to a polypeptide chain; the ribosomes of bacterial cells operate even faster, at a rate of about 20 amino acids per second.

To choreograph the many coordinated movements required for efficient translation, a ribosome contains four binding sites for RNA molecules: one is for the mRNA and three (called the A site, the P site, and the E site) are for tRNAs. A tRNA molecule is held tightly at the A and P sites only if its anticodon forms base pairs with a complementary codon (allowing for wobble) on the mRNA molecule that is threaded through the ribosome. The A and P sites are close enough together for their two tRNA molecules to be forced to form base pairs with adjacent codons on the mRNA molecule. This feature of the ribosome maintains the correct reading frame on the mRNA.

Once protein synthesis has been initiated, each new amino acid is added to the elongating chain in a cycle of reactions containing four major steps: tRNA binding

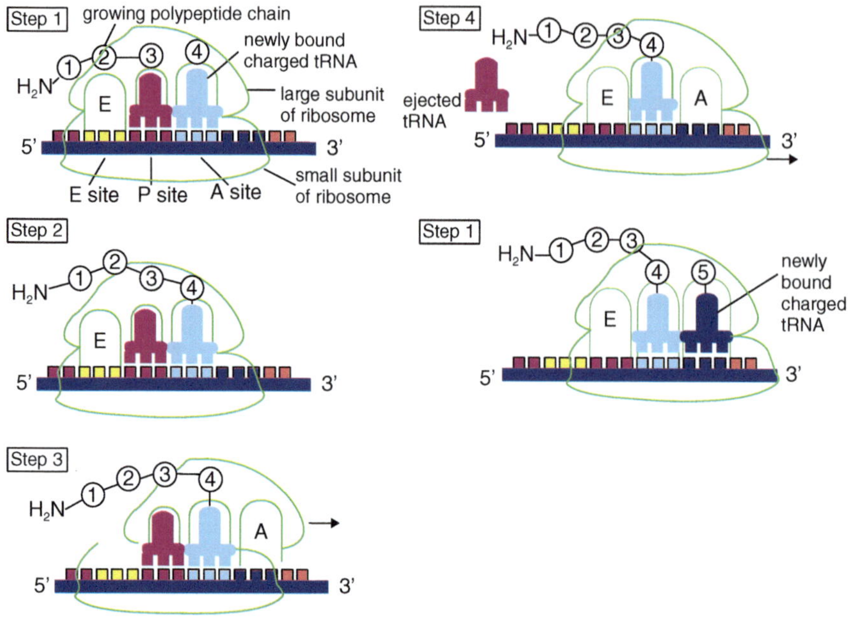

FIGURE 1.18 Translating on the mRNA molecule in a ribosome [1]:343. In a small subunit of the ribosome, it is able to position tRNAs on the mRNA, and in a large subunit of the ribosome, it performs its catalytic activity in forming covalent peptide bonds.

(step 1), peptide bond ([1]:112) formation (step 2), large subunit translocation (step 3), and small subunit translocation (step 4). As a result of the two translocation steps, the entire ribosome moves three nucleotides along the mRNA and is positioned to start the next cycle. Figure 1.18 illustrates this four-step process, beginning at a point at which three amino acids have already been linked together and there is a tRNA molecule in the P site on the ribosome, covalently joined to the C-terminal end of the short polypeptide (a polypeptide is a linear polymer of amino acids linked by peptide bonds; a protein is a large polypeptide [1]:112, Glossary G:25). In step 1, a tRNA carrying the next amino acid in the chain binds to the ribosomal A site by forming base pairs with the mRNA codon positioned there, so that the P site and the A site contain adjacent bound tRNAs. In step 2, the carboxyl end of the polypeptide chain is released from the tRNA at the P site (by breakage of the high-energy bond between the tRNA and its amino acid) and joined to the free amino group of the amino acid linked to the tRNA at the A site, forming a new peptide bond. This central reaction of protein synthesis is catalyzed by a peptidyl transferase ([1]:343) contained in the large ribosomal subunit. In step 3, the large subunit moves to the mRNA held by the small subunit, thereby shifting the acceptor stems of the two tRNAs to the E and P sites of the large subunit. In step 4, another series of conformational changes moves the small subunit and its bound mRNA exactly three nucleotides, ejecting the spent tRNA from the E site and resetting the ribosome so that it is ready to receive the next aminoacyl-tRNA. Step 1 is then repeated with a new incoming aminoacyl-tRNA, and so on.

This four-step cycle is repeated each time an amino acid is added to the polypeptide chain, as the chain grows from its amino end to its carboxyl end [1]:343.

1.6.2 The Ribosome as a Ribozyme

The ribosome is a large complex, which is composed of two thirds RNA and one third protein. In 2000, biologists determined the entire three-dimensional conformation of its large and small subunits. The findings confirm earlier evidence that rRNA – and not proteins – are responsible for the ribosome's overall structure, its ability to position tRNAs on the mRNA, and its catalytic activity in forming covalent peptide bonds. The ribosomal RNAs are folded into highly compact, precise three-dimensional structures that form the compact core of the ribosome and determine its overall shape.

In marked contrast to the central positions of the rRNAs, the ribosomal proteins are generally located on the surface and fill in the gaps and crevices of the folded RNA. Some of these proteins send out extended regions of polypeptide chain that penetrate short distances into holes in the RNA core. The main role of the ribosomal proteins seems to be to stabilize the RNA core, while permitting the changes in rRNA conformation that are necessary for this RNA to catalyze efficient protein synthesis. The proteins also aid in the initial assembly of the rRNAs that make up the core of the ribosome.

Not only the A, P, and E binding sites for tRNAs are formed primarily by ribosomal RNAs, but the catalytic site for peptide bond formation is also formed by RNA, since the nearest amino acid is located more than 1.8 nm away. This discovery came as a surprise to biologists because, unlike proteins, RNA does not contain easily ionizable functional groups that can be used to catalyze sophisticated reactions like peptide bond formation. Moreover, metal ions, which are often used by RNA molecules to catalyze chemical reactions, were not observed at the active site of the ribosome. Instead, it is believed that the 23S rRNA ([1]:346) forms a highly structured pocket that, through a network of hydrogen bonds, precisely orients the two reactions (the growing peptide chain and an aminoacyl-tRNA) and thereby greatly accelerates their covalent joining. An additional surprise came from the discovery that the tRNA in the P site contributes an important OH group to the active site and participates directly in the catalysis. This mechanism may ensure that catalysis occurs only when the P site tRNA is properly positioned in the ribosome.

RNA molecules that possess catalytic activity are known as ribozymes. It is noted that there is good reason to suspect that tRNA rather than protein molecules served as the first catalysts for living cells. If so, the ribosome, with its RNA core, may be a relic of an earlier time in life's history – when protein synthesis evolved in cells that were run almost entirely by ribozymes [1]:347.

1.6.3 The Site in mRNA Initiating Protein Synthesis

The crucial problem is where to start protein synthesis on the mRNA, since the site initiating protein synthesis sets the reading frame for the whole length of the message. An error of one nucleotide either way at this stage would cause every subsequent codon in the message to be misread, resulting in a nonfunctional protein with a garbled sequence of amino acids. The initiation step is also important because for most genes it is the last point at which the cell decides whether the mRNA is to be translated to produce a protein. Thus, it is important where protein synthesis is initiated.

The translation of an mRNA begins with the codon AUG, and a special tRNA is required to start translation. This initiator tRNA always carries the amino acid methionine (in bacteria, a modified form of methionine – formylmethionine – is used), with the result that all newly made proteins have methionine as the first amino acid at their N-terminus, the end of a protein that is synthesized first. (This methionine is usually removed later by a specific protease.) The initiator tRNA is specially recognized by initiation factors because it has a nucleotide sequence distinct from that of the tRNA that normally carries methionine [1]:347. In eukaryotes, the initiator tRNA – methionine complex (Met-tRNAi) – is first loaded into the small ribosomal subunit along with additional proteins called eukaryotic initiation factors (eIFs). Of all the aminoacyl tRNAs in the cell, only the methionine-charged initiator tRNA is capable of tightly binding the small ribosome subunit without the complete ribosome being present, and unlike other tRNAs it binds directly to the P site. That is, the initiator tRNA is specially recognized by initiation factors. The initiator tRNA carries methionine. Translation begins at the first AUG encountered by the small subunit.

Next, the small ribosomal subunit binds to the 5' end of an mRNA molecule, which is recognized by virtue of its 5' cap that has previously bound two initiation factors, eIF4E and eIF4G ([1], Figure 6.38:327). The small ribosomal subunit then moves forward (5' to 3') along the mRNA, searching for the first AUG; additional initiation factors that act as ATP-powered helicases, facilitate this movement. In 90% of mRNAs, translation begins at the first AUG encountered by the small subunit. At this point, the additional initiation factors dissociate, allowing the large ribosomal subunit to assemble with the complex and complete the ribosome. The initiator tRNA remains at the P site, leaving the A site vacant. Protein synthesis is therefore ready to begin ([1], Figure 6.70:348). That is, the small ribosomal subunit binds to the 5' end of an mRNA molecule, where its 5' cap has bound two initiation factors, eIF4E and eIF4G. The small ribosomal subunit moves by energy from additional initiation factors, searching for the first AUG. The large ribosomal subunit binds the small subunit. At this point, translation begins at the first AUG and the initiator tRNA remains at the P site, leaving the A site vacant, ready to begin protein synthesis.

The nucleotides immediately surrounding the start site in eukaryotic mRNAs influence the efficiency of AUG recognition during the foregoing scanning process. If this recognition site differs substantially from the consensus recognition sequence (5'-ACCAUGG-3'), scanning ribosomal subunits will sometimes ignore the first AUG codon in the mRNA and skip to the second or third AUG codon instead. Cells frequently use this phenomenon, known as "leaky scanning," to produce two or more proteins, differing in their N-termini, from the same mRNA molecule. This mechanism allows some genes to produce the same protein with and without a signal sequence attached at its N-terminus, for example, so that the protein is directed to two different compartments in the cell. The mechanism for selecting a start codon in bacteria is different.

Bacterial mRNAs have no 5' caps to signal the ribosome where to begin searching for the start of translation. Instead, each bacterial mRNA contains a specific ribosome-binding site (called the Shine-Dalgarno sequence, named after its discoverers) that is located a few nucleotides upstream of the AUG at which translation is to begin. This nucleotide sequence, with the consensus 5'-AGGAGGU-3', forms base pairs with the 16SrRNA of the small ribosomal subunit to position the initiating AUG codon in the ribosome. A set of translation initiation factors orchestrates this interaction, as well as the subsequent assembly of the large ribosomal subunit to complete the ribosome.

Unlike a eukaryotic ribosome, a bacterial ribosome can readily assemble directly on a ribosome-binding site that lies in the interior of an mRNA molecule, and precedes AUG by several nucleotides. As a result, bacterial mRNAs are often polycistronic: that is, they encode several different proteins, each of which is translated from the mRNA molecule ([1], Figure 6.71:349). By contrast, a eukaryotic mRNA generally encodes only a single protein, or more accurately, a single set of closely related proteins [1]:348.

E1.1 DNA DISCOVERY

Erwin Chargaff's realization that base pairs A-T and C-G were formed, combined with crucially important X-ray crystallography work by Rosalind Franklin and Maurice Wilkins, contributed to James Watson and Francis Crick's derivation of the three-dimensional, double-helical model for the structure of DNA in 1953. The structure immediately revealed how DNA might be copied or duplicated, and it provided the first clues about how a molecule of DNA might encode the instructions for making proteins. The discovery of DNA constructed the platform for understanding life. All genes of global cells consist of parts of base arrangement in DNA. DNA forms core of a chromosome in a cell [2]. In 1962, Watson, Crick, and Wilkins were awarded the Nobel Prize in Physiology and Medicine, for their discoveries concerning the molecular structure of nucleic acids and its significance for information transfer in living organism.

E1.2 MINIMUM RESOLVABLE DISTANCE BY MICROSCOPE

Minimum resolvable by microscope is shown in Figure E1.1. Bacteria can be observed by light microscope, but viruses cannot be observed by it. An electron microscope is necessary to observe viruses.

E1.3 STRUCTURE OF A MICROSCOPE

Figure E1.2 shows the structures of a light microscope and an electron microscope.

E1.4 EUKARYOTES AND PROKARYOTES

All living things (or organisms) are built from cells: small, membrane-enclosed units that are filled with a concentrated aqueous solution of chemicals and endowed with

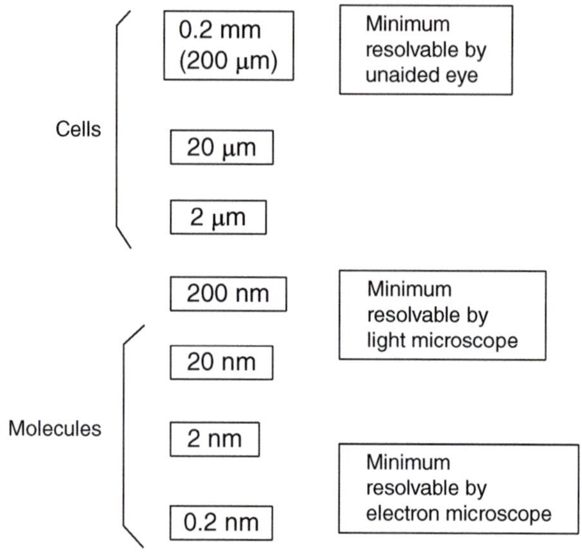

FIGURE E1.1 Minimum resolvable distance by microscope. Electron microscope is necessary to observe virus.

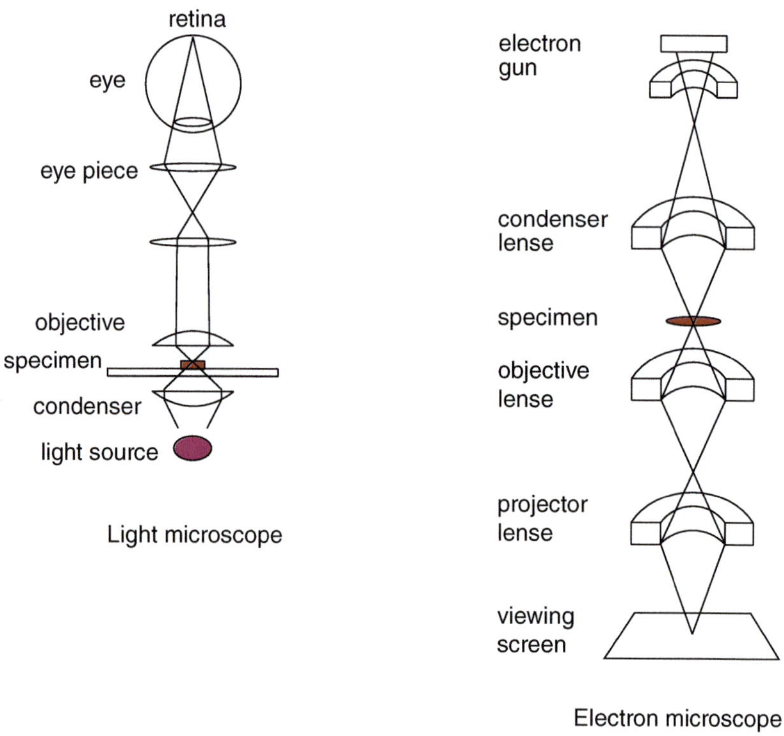

FIGURE E1.2 Structure of a microscope.

the extraordinary ability to create copies of themselves by growing and then dividing into two. Cells, therefore, are the fundamental units of life.

Living things (or organisms) have something in common related to growth, reproduction, and an ability to respond to the environment. Although the cells of all living things are stunningly varied when viewed from the outside, they are fundamentally similar inside. In all organisms, genetic information – in the form of genes – is duplicated in the same way when an organism reproduces. The flow of information, from DNA to mRNA to protein, is so fundamental to life that is referred to as the central dogma.

A cell reproduces by duplicating its DNA and then dividing into two, passing a copy of the genetic instructions encoded in its DNA to each of its daughter cells. The copying, however, is not always perfect, and the instructions are occasionally corrupted by mutations that change the DNA. For this reason, daughter cells do not always match the parent cell exactly.

These simple principles of genetic change and natural selection, applied repeatedly over billions of cell generations, are the basis of evolution – the process by which living species become gradually modified and adapted to their environment in more and more sophisticated ways. Evolution offers an explanation of why present-day cells are so similar in their fundamentals: they have all inherited their genetic instructions from the same common ancestor. It is estimated that this ancestral cell existed between 3.5 and 3.8 billion years ago, and we must suppose that it contained a prototype of the universal machinery of life on Earth today.

Of all the types of cells revealed by microscope (E1.2, E1.3), bacteria have the simplest structure. A bacterium contains no organelles – not even a nucleus to hold its DNA; and this property, the presence or absence of a nucleus, is used as the basis for a simple but fundamental classification of all living things. Organisms whose cells have a nucleus are called "eukaryotes" (from the Greek words "eu," meaning "well" or "truly," and "karyon," meaning a "kernel" or "nucleus"). Organisms whose cells do not have a nucleus are "prokaryotes" (from the Greek "pro," meaning "before") [2]:12.

Prokaryotes often have a tough protective coat, or cell wall, surrounding the plasma membrane, which encloses a single compartment containing the cytoplasm and the DNA.

For example, a virus is a compact package of genetic information in the form of DNA or RNA encased in protein, but it has no ability to reproduce itself by its own effort. Instead, it gets copied by parasitizing the reproductive machinery of the cells that it invades.

All prokaryotes are classified into two distinct domains called the "bacteria" and the "archaea." The species that live in the soil or make us ill are bacteria.

Archaea are found not only in these habitats, but also in environments that are hostile for most other cells, such as concentrated brine, the hot acid of volcanic springs, the airless depths of marine sediments, pools beneath the frozen surface of Antarctica, and in the acidic, oxygen-free environment of a cow's stomach where they break down cellulose and generate methane gas. Many of these extreme environments resemble the harsh conditions that must have existed on the primitive earth, where organisms first evolved before the atmosphere became rich in oxygen.

E1.5 MITOCHONDRION

To maintain their high degree of organization, cells have a constant need for a plentiful supply of ATP. In eukaryotic cells, most of the ATP (Figure 1.17) powering life processes is produced by energy-converting organelles enclosed in a membrane. These are of two types, one of which is mitochondria and the other of which is chloroplasts. Mitochondria, which occur in virtually all cells of animals, plants, and fungi, burn food molecules to produce ATP by oxidative phosphorylation. Chloroplasts, which occur only in plants and green algae, harness solar energy to produce ATP by photosynthesis (Figure E1.3). The features of both mitochondria and chloroplasts are their extensive internal membrane systems, containing sets of membrane protein complexes that work together to produce most of the cell's ATP.

The common evolutionary origin of the energy-converting machinery in mitochondria, chloroplasts, and prokaryotes (archaea and bacteria) is reflected in the fundamental mechanism that they share for harnessing energy. This is known as chemiosmotic coupling, signifying a link between the chemical bond-forming reactions that generate ATP (chemi) and membrane transport processes (osmotic). The chemiosmotic process occurs in two linked stages, both of which are performed by protein complexes in a membrane.

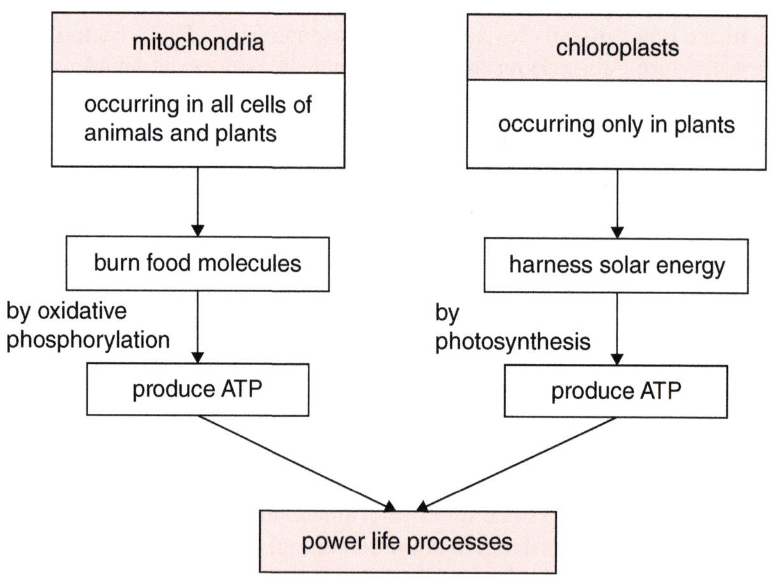

FIGURE E1.3 ATP production in mitochondria and chloroplasts. In eukaryotic cells, most of the ATP powering life processes is produced by energy-converting organelles enclosed in a membrane. These are of two types, one of which is mitochondria and the other of which is chloroplasts. Mitochondria, which occur in virtually all cells of animals, plants, and fungi, burn food molecules to produce ATP by oxidative phosphorylation. Chloroplasts, which occur only in plants and green algae, harness solar energy to produce ATP by photosynthesis.

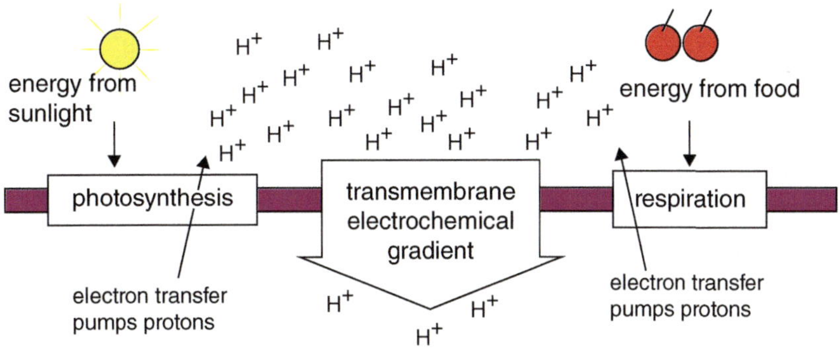

FIGURE E1.4 Stage 1 of chemiosmotic coupling; electron transfer pumps protons [1]:754.

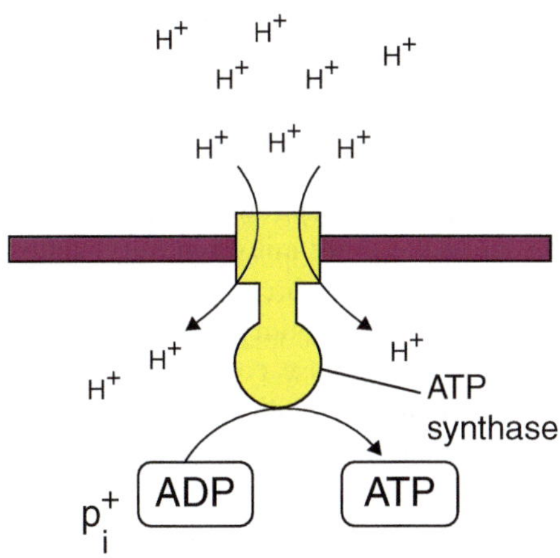

FIGURE E1.5 Stage 2 of chemiosmotic coupling; protons flowback produces ATP from ADP and p_i[1]:754.

Stage 1: High-energy electrons (derived from the oxidation of food, or from pigments excited by sunlight) are transferred along a series of electron-transport protein complexes forming an electron-transport chain embedded in a membrane. Each electron transfer releases a small amount of energy that is used to pump protons (H^+) and thereby generates a large electrochemical gradient across the membrane (Figure E1.4). Such an electrochemical gradient provides a way of storing energy, and it can be harnessed to do useful work when ions flow back across the membrane.

Stage 2: The protons flow back down their electrochemical gradient through a membrane protein machine called "ATP synthase," which catalyzes the production of ATP from ADP and inorganic phosphate (p_i). This enzyme works like a turbine in the membrane, driven by protons, in order to synthesize ATP (Figure E1.5). In this way, the energy derived from food or sunlight in stage 1 is converted into the chemical energy of a phosphate bond in ATP [1]:754.

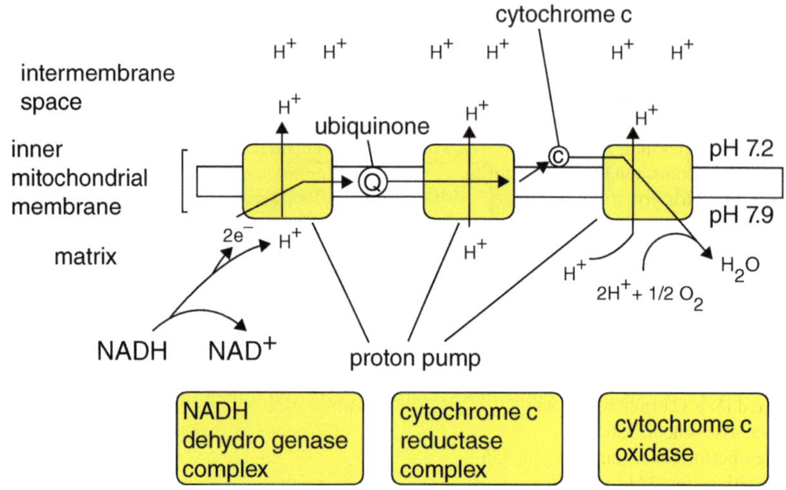

FIGURE E1.6 Electron-transport chain pumping protons in mitochondrion [2]:456.

The electron-transport chain, or respiratory chain, which carries out oxidative phosphorylation is present in the inner mitochondrial membrane. Each chain contains over 40 proteins, grouped into three large respiratory enzyme complexes: (1) NADH (nicotinamide adenine dinucleotide) dehydrogenase complex: NADH→$2e^-$+ NAD^+ + H^+, (2) cytochrome c reductase complex, and (3) cytochrome c oxidase complex (Figure E1.6). Each complex facilitates the high-energy electrons' passage. Ubiquinone (Q) and cytochrome c (c) serve as mobile carriers that ferry electrons from one complex to the next. The movement of electrons through these respiratory complexes is accompanied by the pumping of protons from the mitochondrial matrix to the intermembrane space. Each complex can be thought of as a proton pump. The proton pump generates a H^+ gradient across the inner membrane, resulting in about 0.7 units higher pH in the matrix (around 7.9) than the intermembrane space (7.2, the same pH as the cytosol). The final reaction is the only oxygen-requiring step in cell respiration, and it consumes nearly all of the oxygen that we breathe [2]:455–456.

2 The Cell Cycle

The only way to produce a new cell is to duplicate an already existing cell. This simple fact, first established in the middle of the nineteenth century, carries with it a profound message for the continuity of life. Repeated rounds of cell growth and division have produced all living organisms, from the unicellular bacterium to the multicellular mammal, extending back in time to the beginning of life on Earth over three billion years ago.

A cell reproduces by performing an orderly sequence of events in which it duplicates its contents and then divides into two. This cycle of duplication and division, known as the cell cycle, is the essential mechanism by which all living things reproduce. In unicellular species, such as bacteria and yeasts, each cell division produces a complete new organism. In multicellular species, long and complex sequences of cell divisions are required to produce a functioning organism. Even in the adult body, cell division is usually needed to replace cells that die. In fact, each of us must manufacture many millions of cells every second simply to survive: if all cell division were stopped – by exposure to a very large dose of x-rays, for example – we would die within a few days. The details of the cell cycle vary from organism to organism and at different times in an organism's life. Certain characteristics, however, are universal. At least, the cell must perform its most fundamental task: passing on its genetic information to the next generation of cells. The DNA in each chromosome must first be faithfully replicated to produce two complete copies: two genetically identical daughter cells. The replicated chromosomes must then be accurately distributed (segregated) to the two daughter cells, so that each receives a copy of the entire genome (Figure 2.1).

In addition to duplicating their genome, most cells also duplicate their organelles and macromolecules; otherwise, daughter cells would get smaller with each division. To maintain their size, dividing cells must coordinate their growth (that is, their increase in cell mass) with their division.

This chapter describes the events of the cell cycle and how they are controlled and coordinated. We begin with a brief overview of the cell cycle. We then describe the cell cycle control system, a complex network of regulatory proteins that triggers the different events of the cycle. We next consider in detail the major stages of the cell cycle, in which the chromosomes are duplicated and then segregated into the two daughter cells. Finally, we consider how extracellular signals govern the rates of cell growth and division and how these two processes are coordinated.

2.1 THE FOUR PHASES OF THE CELL CYCLE

2.1.1 Overview of the Cell Cycle

The most basic function of the cell cycle is to duplicate the vast amount of DNA in the chromosomes and then segregate the copies into two genetically identical

DOI: 10.1201/9781003780663-2

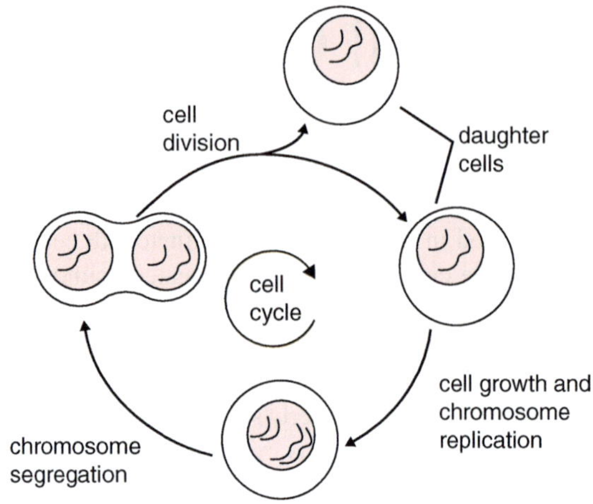

FIGURE 2.1 The cell cycle [1]:964. The division of a hypothetical eukaryotic cell with two chromosomes is shown to illustrate how two genetically identical daughter cells are produced in each cycle. Each of the daughter cells will often continue to divide by going through additional cell cycles.

daughter cells. These processes define the two major phases of the cell cycle: the S phase and the M phase. Chromosome duplication occurs during the S phase (S for DNA synthesis), which requires 10–12 hours and occupies about half of the cell cycle time in a typical mammalian cell. After the S phase, chromosome segregation and cell division occur in the M phase (M for mitosis), which requires much less time (less than an hour in a mammalian cell). The M phase comprises two major events: nuclear division, or mitosis, during which the copied chromosomes are distributed into a pair of daughter nuclei; and cytoplasmic division, or cytokinesis, when the cell itself divides into two. That is, the major events of the cell cycle are as follows. The chromosomal duplication occurs in the S phase, and in the M phase, the duplicated chromosomes are segregated into a pair of daughter nuclei (in mitosis), after which the cell itself divides into two (cytokinesis).

2.1.2 The Four Phases of the Cell Cycle

Most cells require much more time to grow and double their mass of proteins and organelles than they require to duplicate their chromosomes and divide. Partly to allow time for growth, most cell cycles have gap phases – a G_1 phase between the M phase and the S phase and a G_2 phase between the S phase and the M phase. Thus, the eukaryotic cell cycle is traditionally divided into four sequential phases: G_1, S, G_2, and M. G_1, S, and G_2 together are called interphase (Figure 2.2). In a typical human cell proliferating in culture, interface might occupy 23 hours of a 24-hour cycle, with 1 hour for the M phase. Cell growth occurs throughout the cell cycle, except during mitosis.

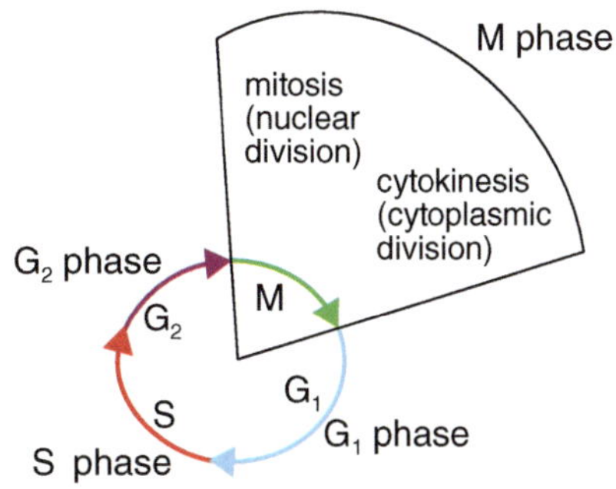

FIGURE 2.2 The four phases of the cell cycle [1]:965. In most cells, gap phases separate the major events of the S phase and the M phase. G_1 is the gap between the M phase and the S phase, while G_2 is the gap between the S phase and the M phase.

The two gap phases require more time than the simple time for duplication and division to allow cell growth. They also provide time for the cell to monitor the internal and external environments to ensure that conditions are suitable and preparations are complete before the cell commits itself to the major upheavals of synthesis and mitosis. The G_1 phase is especially important in this respect. Its length can vary greatly depending on external conditions and extracellular signals from other cells. If extracellular conditions are unfavorable, for example, cells delay progress through G_1 and may even enter a specialized resting state known as G_0 (G zero), in which they can remain for days, weeks, or even years before resuming proliferation. Indeed, many cells remain permanently in G_0 until they or the organism dies.

If extracellular conditions are favorable and signals to grow and divide are present, cells in early G_1 or G_0 progress through a commitment point near the end of G_1 known as Start (in yeasts) or the restriction point (in mammalian cells). We will use the term Start for both yeast and animal cells. After passing this point, cells are committed to DNA replication, even if extracellular signals that stimulate cell growth and division are removed [1]:965.

2.1.3 The Cell Cycle Control System

The cell cycle control system operates to trigger the events of the cell cycle in a set sequence (Figure 2.3). In its simplest form – as seen in the stripped-down cell cycles of early animal embryos, for example – the control system is rigidly programmed to provide a fixed amount of time for the completion of each cell cycle event. The control system in these early embryonic divisions is independent of the events it controls, so that its timing mechanisms continue to operate even if those events fail. However, in most cells, the control system does respond to information received back from the processes it controls. If some malfunction prevents the successful completion of DNA synthesis, for example, signals are sent to the control system to delay progression to the M phase. Such delays provide time for the machinery to be repaired and also prevent the disaster that might result if the cycle progressed prematurely to the next stage – and segregated incompletely replicated chromosomes, for example.

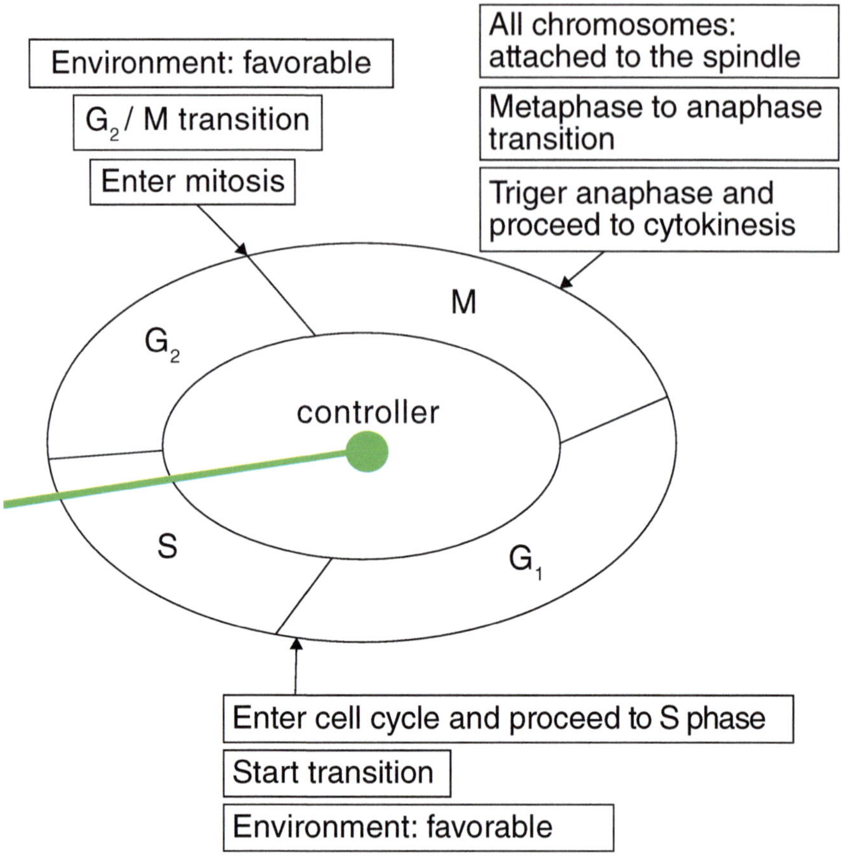

FIGURE 2.3 The control of the cell cycle [1]:968. A cell cycle control system triggers the essential processes of the cycle – such as the DNA replication, mitosis, and cytokinesis. The control system is represented here as a central arm – the controller – that rotates clockwise, triggering essential processes when it reaches specific transition on the outer dial. Information about the completion of cell cycle events, as well as signals from the environment, can cause the control system to arrest the cycle at these transitions.

The cell cycle control system is based on a connected series of biochemical switches, each of which initiates a specific cell cycle event. This system of switches possesses many important features that increase the accuracy and reliability of cell cycle progression. Firstly, the switches are generally binary (on/off) and launch events in a complete, irreversible fashion. It would clearly be disastrous, for example, if events like chromosome condensation or nuclear-envelope breakdown were only partially initiated or started but not completed. Secondly, the cell cycle control system is remarkably robust and reliable partly because backup mechanisms and other features allow the system to operate effectively under a variety of conditions and even if some components fail. Finally, the control system is highly adaptable and can be modified to suit specific cell types or to respond to specific intracellular or extracellular signals [1]:967.

In most eukaryotic cells, the cell cycle control system administers cell cycle progression at three major regulatory transitions (Figure 2.3). The first is Start (or the restriction point) in late G_1, when the cell commits to cell cycle entry and chromosome duplication. The second is the G_2/M transition, when the control system triggers the early mitosis events that lead to chromosome alignment on the mitotic spindle in metaphase. The third is the metaphase-to-anaphase transition, when the control system stimulates sister-chromatid separation, leading to the completion of mitosis and cytokinesis. The control system blocks progression through each of these transitions if it detects problems inside or outside the cell. If the control system senses problems in the completion of DNA replication, for example, it will hold the cell at the G_2/M transition until those problems are solved. Similarly, if extracellular conditions are not appropriate for cell proliferation, the control system blocks progression through Start, thereby preventing cell division until conditions become favorable [1]:968. That is, at Start and the G_2/M transition, the control system blocks progression if extracellular conditions are not appropriate for cell proliferation.

2.2 CYCLIN-CDK AS CENTRAL COMPONENTS OF THE CELL CYCLE

2.2.1 Cyclin-Dependent Protein Kinase (CDK)

Members of a family of protein kinases known as cyclin-dependent kinases (CDKs) are central components of the cell cycle control system. Changes in the activities of these kinases during the cycle lead to cyclical changes in the phosphorylation of intracellular proteins that initiate or regulate the major events of the cell cycle. An increase in CDK activity at the G_2/M transition, for example, increases the phosphorylation of intracellular proteins that initiate or regulate the major events of the cell cycle. An increase in CDK activity at the G_2/M transition, for example, increases the phosphorylation of proteins that control chromosome condensation, nuclear-envelope breakdown, spindle assembly, and other events that occur in early mitosis.

Cyclical changes in CDK activity are controlled by a complex array of enzymes and other proteins. The most important of these CDK regulators are proteins known as cyclins. CDKs, as their name implies, are dependent on cyclins for their activity: unless they are bound tightly to a cyclin, they have no protein kinase activity. Cyclical changes in cyclin protein levels result in the cyclic assembly and activation of cyclin-CDK complex at specific stages of the cell cycle.

There are four classes of cyclins, each defined by the stage of the cell cycle at which they bind CDKs and function. All eukaryotic cells require three of these classes.

1. G_1/S-cyclins activate CDKs in late G_1 and thereby help trigger progression through Start, resulting in a commitment to cell cycle entry. Their levels fall in the S phase.

2. S-cyclins bind CDKs soon after progression through Start and help stimulate chromosome duplication. S-cyclin levels remain elevated until mitosis, and these cyclins also contribute to the control of some early mitosis events.
3. M-cyclins activate CDKs that stimulate entry into mitosis at the G_2/M transition. M-cyclin levels fall in mid-mitosis.

In most cells, the fourth class of cyclins, the G_1-cyclins, helps to govern the activities of the G_1/S-cyclins, which control progression through Start in late G_1.

In yeast cells, a single CDK protein binds all classes of cyclins and triggers different cell cycle events by changing cyclin partners at different stages of the cycle. In vertebrate cells, by contrast, there are four CDKs. Two interact with G_1-cyclins, one with G_1/S and S-cyclins, and one with S and M-cyclins. We simply refer to the different cyclin-CDK complexes as G_1-CDK, G_1/S-CDK, S-CDK, and M-CDK.

Different cyclin-CDK complexes trigger different cell cycle events as follows. The cyclin protein does not simply activate its CDK partner but also direct it to specific target proteins. As a result, each cyclin-CDK complex phosphorylates a different set of substrate proteins. The same cyclin-CDK complex can also induce different effects at different times in the cycle, probably because the accessibility of some CDK substrates changes during the cell cycle. Certain proteins that function in mitosis, for example, may become available for phosphorylation only in G_2 (Figure 2.6 and 2.7, [1]:969).

2.2.2 The Regulation of CDK Activity

In this subsection, we describe what is the determinant of CDK activity. The rise and fall of the cyclin level is the primary determinant of CDK activity during the cell cycle. Several additional mechanisms, however, help control CDK activity at specific stages of the cycle.

Phosphorylation at a pair of amino acids in the roof of the kinase active site inhibits the activity of a cyclin-CDK complex. Phosphorylation of these sites by a protein kinase known as Weel inhibits CDK activity, while dephosphorylation of these sites by a phosphatase known as Cdc25 increases CDK activity (Figure 2.4). This regulatory mechanism is particularly important in the control of M-CDK activity at the onset of mitosis.

The binding of CDK inhibitor proteins (CKIs) inactivates cyclin-CDK complexes. The three-dimensional structure of a cyclin-CDK-CKI complex reveals that CKI binding stimulates a large rearrangement in the structure of the CDK active site, rendering it inactive (Figure 2.5). Cells use CKIs primarily to help govern the activities of G_2/S and S-CDKs early in the cell cycle.

2.2.3 Regulated Proteolysis Triggering the Metaphase-To-Anaphase

In this subsection, we describe that the key regulator of the metaphase-to-anaphase transition is the enzyme APC/C. Whereas activation of specific cyclin CDK complexes drives progression through the Start and G_2/M transitions, progression through

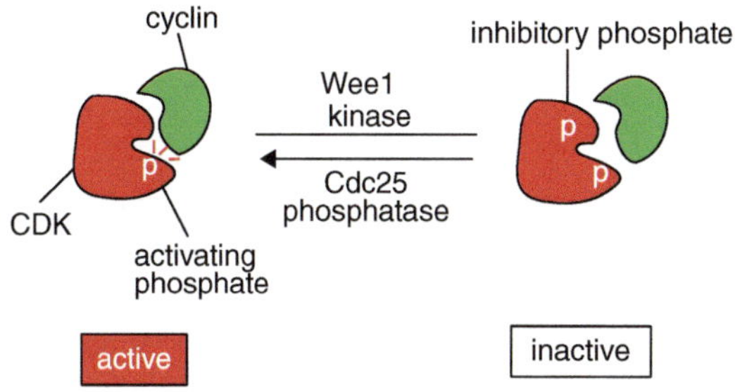

FIGURE 2.4 The regulation of CDK activity by phosphorylation [1]:970. The active cyclin-CDK complex is turned off when the kinase Weel phosphorylates two closely spaced sites above the active site. Removal of these phosphates by the phosphatase Cdc25 activates the cyclin-CDK complex. CDK-activating kinase (CAK) adds the activating phosphate.

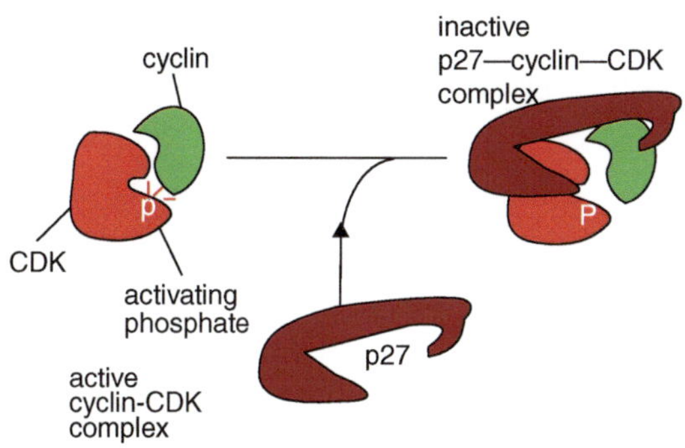

FIGURE 2.5 The inhibition of a cyclin-CDK complex by a CKI [1]:971. The p27 binds to both the cyclin and CDK in the complex, distorting the active site of the CDK.

the metaphase-to-anaphase transition is triggered not by protein phosphorylation but by protein destruction, leading to the final stages of cell division.

The key regulator of the metaphase-to-anaphase transition is the anaphase-promoting complex, or cyclosome (APC/C), a member of the ubiquitin ligase family of enzymes. These enzymes are used in numerous cell processes to stimulate the proteolytic destruction of specific regulatory proteins. They polyubiquitylate specific target proteins, resulting in their destruction in proteasome.

The APC/C catalyzes the ubiquitylation and destruction of two major types of proteins. The first is securin (see Figure 2.11), which protects the protein linkages that hold sister-chromatid pairs together in early mitosis. Destruction of securin in metaphase activates a protease called separase ([1]:993) that separates the sisters and unleashes anaphase, resulting in chromatid-separation. The S- and M-cyclins are the second major targets of the APC/C (Figure 2.6). Destroying these cyclins inactivates most CDKs in the cell, leading to the many proteins phosphorylated by CDKs from the S phase to early mitosis being dephosphorylated by various phosphatases in the anaphase cell, and leading to chromatid separation and cell division.

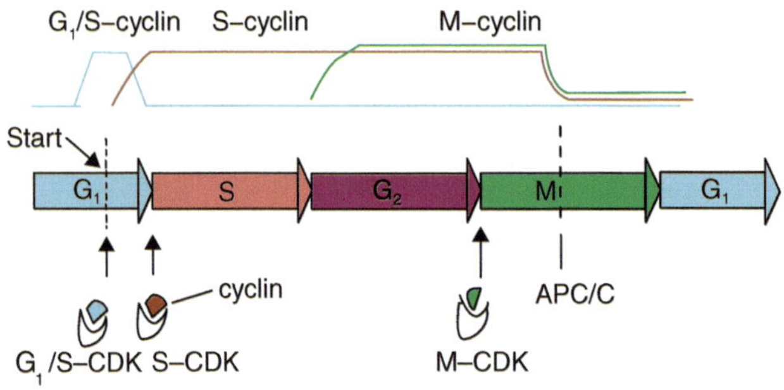

FIGURE 2.6 Cyclin-oscillation during the cell cycle [1]:969. The concentrations of the three major cyclin types oscillate during the cell cycle. In late G_1, rising G_1/S-cyclin levels lead to the formation of G_1/S-CDK complexes that trigger progression through the Start transition. S-CDK complexes form at the start of the S phase and trigger DNA replication, and S-cyclin levels are held high until early mitosis. M-CDK complexes form during G_2 but are held in an inactive state; they are activated at the end of G_2 and trigger entry into mitosis at the G_2/M transition. A separate regulatory protein complex, the APC/C, initiates the metaphase-to-anaphase transition [1]:969.

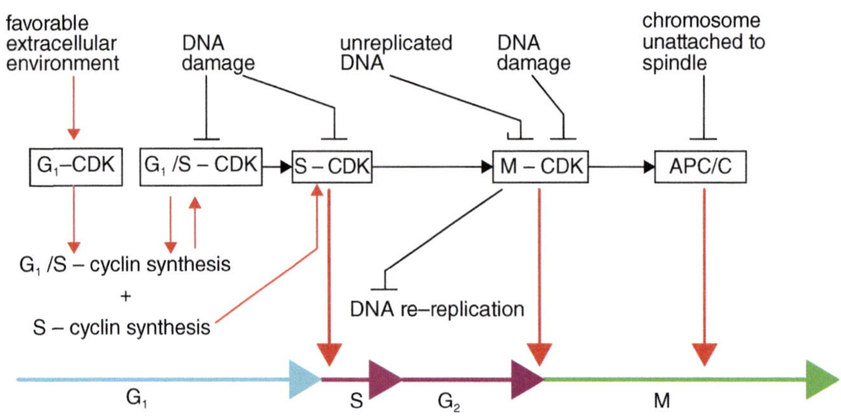

FIGURE 2.7 An overview of the cell cycle control system [1]:973. The core of the cell cycle control system consists of a series of cyclin-CDK complexes (in rectangle). The activity of each complex is also influenced by various inhibitory mechanisms, which provide information about the extracellular environment, cell damage, and incomplete cell cycle events (top). The APC/C catalyzes the ubiquitylation and consequent destruction of specific regulatory proteins that control critical events in the cycle.

This dephosphorylation of CDK targets is needed for the completion of the M phase, including the final steps in mitosis and then cytokinesis. Following its activation in mid-mitosis, the APC/C remains active in G_1 to provide a stable period of CDK inactivity. When G_1/S-CDK is activated in late G_1, the APC/C is turned off, thereby allowing cyclin accumulation in the next cell cycle (Figure 2.7) [1]:971.

2.3 DNA DUPLICATION AND SISTER CHROMATIDS

2.3.1 The S Phase

In this subsection, we describe the remarkably accurate mechanism of DNA duplication. The linear chromosomes of eukaryotic cells are vast and dynamic assemblies of DNA and protein, and their duplication is a complex process that occupies a major fraction of the cell cycle. Not only must the long DNA molecule of each chromosome be duplicated accurately – a remarkable feat in itself – but the protein packaging surrounding each region of the DNA must also be reproduced, ensuring that the daughter cells inherit all features of the chromosome structure.

The control event of chromosome duplication – DNA replication – poses two problems for the cell. Firstly, replication must occur with extreme accuracy to minimize the risk of mutations in the next generation. Secondly, every nucleotide in the genome must be copied once, and only once, to prevent the damaging effects of gene amplification. In this subsection, we consider the elegant mechanisms by which the cell cycle control system initiates the replication process and, at the same time, prevents it from happening more than once per cycle.

DNA replication begins at origins of replication, which are scattered at numerous locations in every chromosome. During the S phase, DNA replication is initiated at these origins when a DNA helicase unwinds the double helix and DNA replication enzymes are located onto the two single-stranded templates. This leads to the elongation phase of replication, when the replication machinery moves outward from the origin at two replication forks.

To ensure that chromosome duplication occurs only once per cell cycle, the initiation phase of DNA replication is divided into two distinct steps that occur at different times in the cell cycle. The first step occurs in late mitosis and early G_1, when a pair of inactive DNA helicases is loaded onto the replication origin, forming a large complex called the prereplicative complex (pre-RC). This step is sometimes called licensing of replication origins because initiation of DNA synthesis is permitted only at origins containing a pre-RC. The second step occurs in the S phase, when the DNA helicases are activated, resulting in DNA unwinding and the initiation of DNA synthesis. Once a replication origin has been fired in this way, the two helicases move out from the origin with the replication forks, and the origin cannot be reused until a new pre-RC is assembled there at the end of mitosis. That is, the chromosome duplication must occur only once per cell cycle. The initiation phase of DNA replication is divided into two steps. In the first step in late mitosis and early G_1, a pair of inactive DNA helicases is loaded onto the replication origin, forming a pre-RC. The initiation of DNA synthesis is permitted only at origins containing a pre-RC. In the second step in the S phase, the DNA helicases are activated, resulting in the initiation of DNA synthesis. Once a replication origin has been fired, the origin cannot be reused until a new pre-RC is assembled there at the end of mitosis. As a result, origins can be activated only once per cell cycle.

Figure 2.8 illustrates some of the molecular details underlying the control of the two steps in the initiation of DNA replication. A key player is a large multiprotein complex called the origin recognition complex (ORC), which binds to replication origins throughout the cell cycle. In late mitosis and early G_1, the proteins Cdc6 and

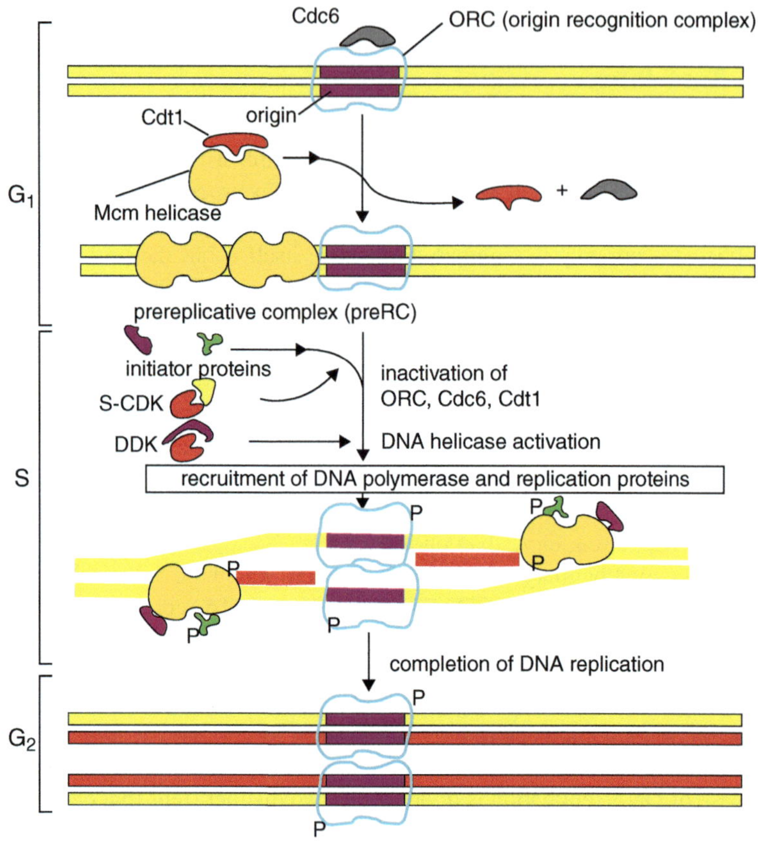

FIGURE 2.8 Control of the initiation of DNA replication [1]:976. The replication origin is bound by ORC throughout the cell cycle. In early G_1, Cdc6 associates with the ORC, and these proteins bind the DNA helicase, which contains six closely related subunits called MCM proteins. The helicase also associates with a protein called Cdt1. Using energy provided by ATP hydrolysis (Figure 1.17), the ORC and Cdc6 proteins load two copies of the DNA helicase, in an inactive form, around the DNA next to the origin, thereby forming the prereplicative complex (pre-RC). At the onset of the S phase, S-CDK stimulates the assembly of several initiator proteins on each DNA helicase, while another protein kinase, DDK, phosphorylates subunits of the DNA helicase. As a result, the DNA helicases are activated and unwind the DNA. DNA polymerase and other replication proteins are recruited to the origin, and DNA replication begins. The ORC is displaced by the replication machinery and then rebinds. S-CDK and other mechanisms also inactivate the pre-RC components ORC, Cdc6, and Cdt1, thereby preventing formation of new pre-RCs at the origins until the end of mitosis.

Cdt1 collaborate with the ORC to load the inactive DNA helicases around the DNA next to the origin. The resulting large complex is the pre-RC and the origin is now licensed for replication.

At the onset of the S phase, S-CDK triggers origin activation by phosphorylating specific initiator proteins, which then nucleate the assembly of a large protein complex that activates the DNA helicase and recruits the DNA synthesis machinery. Another protein kinase called DDK is also activated in the S phase and helps drive origin activation by phosphorylating specific subunits of DNA helicase.

At the same time, as S-CDK initiates DNA replication, several mechanisms prevent assembly of new pre-RCs. S-CDK phosphorylates and thereby inhibits the ORC and Cdc6 proteins. Inactivation of the APC/C in late G_1 also helps turn off pre-RC assembly. That is, initiating DNA replication leads to preventing assembly of a new pre-RC. Inactivation of the APC/C in late G_1 also helps turn off pre-RC assembly.

In late mitosis and early G_1, the APC/C triggers the destruction of an inhibitor called geminin inhibiting Cdt1, thereby allowing Cdt1 to be active. When the APC/C is turned off in late G_1, geminin accumulates and inhibits the Cdt1 that is not associated with DNA. Also, the association of Cdt1 with a protein at active replication forks stimulates Cdt1 destruction. Therefore, in late G_1, it is not possible to prepare the pre-RC assembly for the next replication, while in late mitosis and early G_1, active APC/C and active Cdt1 allow pre-RC assembly to prepare for the next S phase. That is, in late mitosis and early G_1, pre-RC is prepared for the next S phase, by active APC/C and active Cdt1.

In these various ways, pre-RC formation is prevented from S phase to mitosis, thereby ensuring that each origin is fired only once per cell cycle. How, then, is the cell cycle control system reset to allow replication in the next cell cycle? At the end of mitosis, APC/C activation leads to the inactivation of CDKs and the destruction of geminin. ORC and Cdc6 are dephosphorylated and Cdt1 is activated, allowing pre-RC assembly to prepare the cell for the next S phase [1]:976.

2.3.2 Chromosome Duplication Requiring Duplication of Chromatin Structure

In this subsection, we describe that chromatin consists of various protein components, including DNA, histone, and non-histones. The DNA of the chromosomes is extensively packaged in a variety of protein components, including histones and various regulatory proteins involved in the control of gene expression. Thus, duplication of a chromosome is not simply a matter of replicating the DNA at its core but also requires the duplication of these chromatin (Figure 2.9) proteins and their proper assembly on the DNA.

The production of chromatin proteins increases during the S phase to provide the raw materials needed to package the newly synthesized DNA.

Most importantly, S-CDK stimulates a large increase in the synthesis of the four histone subunits that form the histone octamers at the core of each nucleosome [1]:205. These subunits are assembled into nucleosomes on the DNA by nucleosome assembly factors, which typically associate with the replication fork and distribute nucleosomes on both strands of the DNA as they emerge from the DNA synthesis machinery. Chromatin packaging helps to control gene expression. In some parts of the chromosome, the chromatin is highly condensed and is called heterochromatin [1]:205, whereas in other regions it has a more open structure and is called euchromatin [1]:205. These differences in chromatin structure depend on a variety of mechanisms, including modification of histone tails and the presence of non-histone

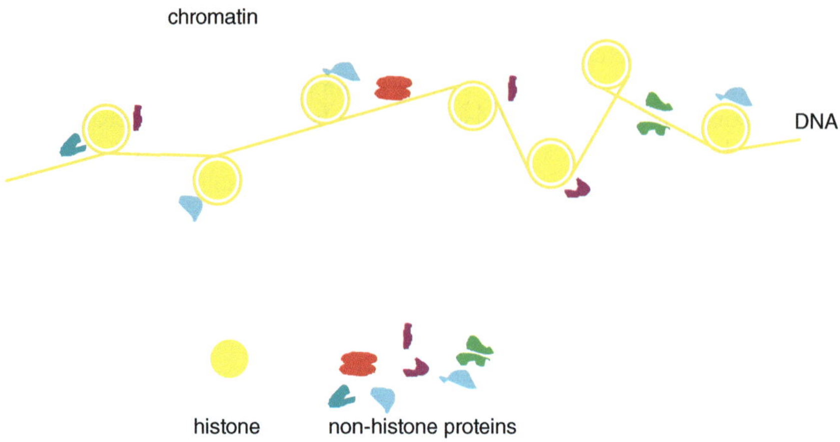

FIGURE 2.9 Chromatin [1]:187. Chromatin consists of DNA bound to both histone and non-histone proteins. The complex of both classes of protein with the nuclear DNA of eukaryotic cells is known as chromatin.

proteins. Because these differences are important in gene regulation, it is crucial that chromatin structure, like the DNA within, is reproduced accurately during the S phase. It is not understood how chromatin structure is reproduced, however. During DNA synthesis, histone-modifying enzymes ([1]:201) and various non-histone proteins are probably deposited onto the two new DNA strands as they emerge from the replication fork, and it is thought for these proteins to reproduce the local chromatin structure of the parent chromosome [1]:975.

2.3.3 Cohesins Holding Sister Chromatids Together

In this subsection, we describe that cohesin is deposited at many locations along the length of each sister chromatid as the DNA is replicated in the S phase [1]:977. At the end of the S phase, each replicated chromosome consists of a pair of identical sister chromatids glued together along their length. This sister-chromatid cohesion sets the stage for a successful mitosis because it greatly facilitates the attachment of the two sister chromatids to opposite poles of the mitotic spindle (Figure 2.11). Imagine how difficult it would be to achieve this bipolar attachment if sister chromatids were allowed to drift apart after the S phase. Indeed, defects in sister-chromatid cohesion – in yeast mutants, for example – lead inevitably to major errors in chromosome segregation.

Sister-chromatid cohesion depends on a large protein complex called cohesin, deposited at many locations along the length of each sister chromatid at the DNA replication in the S phase. Two of the subunits of cohesin are members of a large family of proteins called SMC proteins (for Structural Maintenance of Chromosomes). Cohesin forms giant ringlike structures, and it has been proposed that these surround the two sister chromatids. Cohesins hold sister chromatids together.

Sister-chromatid cohesion also results, at least in part, from DNA catenation, because the intertwining of sister DNA molecules occurs when two replication forks meet during DNA synthesis. The enzyme topoisomerase II ([1]:253) gradually disentangles the catenated sister DNAs between the S phase and early mitosis by cutting one DNA molecule, passing the other through the break, and then resealing the cut DNA ([1]:253). Once the catenation has been removed, sister-chromatid cohesion depends primarily on cohesin complexes. The sudden and synchronous loss of sister cohesion at the metaphase-to-anaphase transition therefore depends primarily on disruption of these complexes (Figure 2.11), leading to sister-chromatids separation.

2.4 MITOSIS

2.4.1 Mitosis

In this subsection, we describe that during mitosis, the sister chromatids are separated and distributed (segregated), and after mitosis, cytokinesis divides the cell into two. Following the completion of the S phase and transition through G_2, the cell undergoes the dramatic upheaval of the M phase. This begins with mitosis, during which the sister chromatids are separated and distributed (segregated) to a pair of identical daughter nuclei, each with its own copy of the genome. Mitosis is traditionally divided into five stages – prophase, prometaphase, metaphase, anaphase, and telophase – defined primarily on the basis of chromosome behavior as seen under a microscope. As mitosis is completed, the second major event of the M phase – cytokinesis – divides the cell into two halves, each with an identical nucleus. That is, during mitosis, the sister chromatids are separated and segregated into a pair of identical daughter nuclei. At the completion of mitosis, the second event of the M phase, cytokinesis, results in the cell's division into two, each with identical nucleus.

Mitosis can be divided into two major parts, each governed by distinct components of the cell cycle control system. Firstly, an abrupt increase in M-CDK activity at the G_2/M transition triggers the events of early mitosis (prophase, prometaphase, and metaphase). During this period, M-CDK and several other mitotic protein kinases phosphorylate a variety of proteins, leading to the assembly of the mitotic spindle and its attachment to the sister-chromatid pairs. The second major part of mitosis begins sister-chromatid separation at the metaphase-to-anaphase transition, when the APC/C triggers the destruction of securin (Figure 2.11), liberating a protease called separase that clears cohesin and thereby initiates separation of the sister chromatids. The APC/C also promotes the destruction of cyclins, which leads to CDK inactivation and the dephosphorylation of CDK targets, which is required for all events of the late M phase, including the completion of anaphase, the disassembly of the mitotic spindle, and the division of the cell by cytokinesis. That is, in the late M phase, cell division is completed.

In the following, we describe the key mechanical events of mitosis and how M-CDK and the APC/C orchestrate them [1]:978.

2.4.2 M-CDK Driving Entry into Mitosis

In this subsection, we describe that M-CDK induces the assembly of the mitotic spindle, chromosome condensation, the reorganization of intertwined sister chromatids, and the breakdown of the nuclear envelope. One of the most remarkable features of cell cycle control is that a single protein kinase, M-CDK, gives rise to all of the diverse and complex cell rearrangements that occur in the early stages of mitosis. At least, M-CDK must induce the assembly of the mitotic spindle and ensure that each sister chromatid in a pair is attached to the opposite pole of the spindle. It also triggers chromosome condensation, the large-scale reorganization of the intertwined sister chromatids into compact, rodlike structures. In animal cells, M-CDK also promotes the breakdown of the nuclear envelope and rearrangements of the actin cytoskeleton and the Golgi apparatus. Each of these processes is thought to be initiated when M-CDK phosphorylates specific proteins involved in the process, although most of these proteins have not yet been identified.

M-CDK does not act alone to phosphorylate key proteins involved in early mitosis. Two additional families of protein kinases, the Polo-like kinase and the Aurora kinases, also make important contributions to the control of early mitotic events. The Polo-like kinase (Plk), for example, is required for the normal assembly of a bipolar mitotic spindle, in part because it phosphorylates proteins involved in separation of the spindle poles early in mitosis. The Aurora kinase Aurora-A also helps control proteins that govern the assembly and stability of the spindle, whereas Aurora-B controls attachment of sister chromatids to the spindle [1]:978. Thus, in early mitosis, two additional protein kinases, the Polo-like kinase and the Aurora kinases, make an important contribution to early mitotic events; the normal assembly of a bipolar mitotic spindle, the assembly and stability of the spindle, and attachment of sister chromatids to the spindle.

2.4.3 Dephosphorylation Activating M-CDK at the Onset of Mitosis

In this subsection, we describe what triggers the activation of the M-CDK in mitosis. M-CDK activation begins with the accumulation of M-cyclin. In embryonic cell cycles, the synthesis of M-cyclin is constant throughout the cell cycle, and M-cyclin accumulation results from the high stability of the protein in interphase. However, in most cell types, M-cyclin synthesis increases during G_2 and M, owing primarily to an increase in M-cyclin gene transcription. The increase in M-cyclin protein leads to a corresponding accumulation of M-CDK (the complex of Cdk1 and M-cyclin) as the cell approaches mitosis. Although the CDK in these complexes is phosphorylated at an activating site by the CDK-activating kinase (CAK), the protein kinase Weel holds it in an inactive state by inhibitory phosphorylation at two neighboring sites in CDK (Figure 2.10). Thus, by the time the cell reaches the end of G_2, it contains an abundant stockpile of M-CDK that is primed and ready to act but is suppressed by phosphates that block the active site of the kinase. That is, the increase of M-cyclin synthesis during G_2 and the M phase leads to an accumulation of M-CDK (the complex of Cdk1 and M-cyclin).

The M-CDK stockpile is activated by the active protein phosphatase Cdc25, which removes the inhibitory phosphates that restrain M-CDK (Figure 2.10). At the same time, the inhibitory activity of the kinase Weel is suppressed, further ensuring that M-CDK activity increases. The mechanisms that unleash Cdc25 activity in early mitosis are not well understood. One possibility is that the M-CDKs, which are active in G_2 and early prophase (early mitosis), stimulate Cdc25 ([1]:979). That is, in

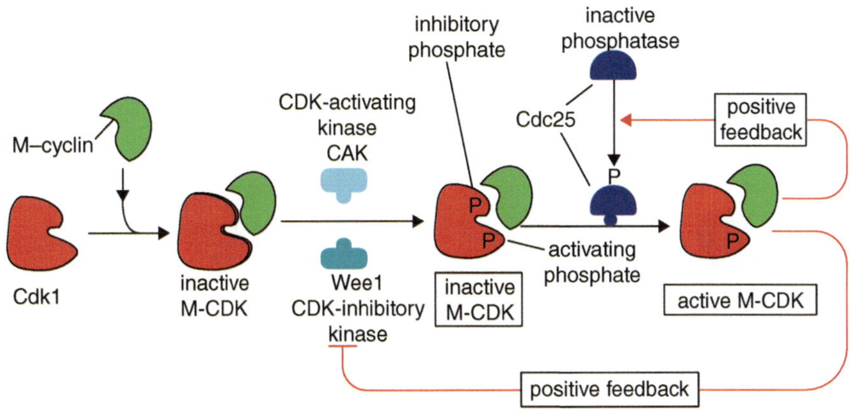

FIGURE 2.10 The activation of M-CDK [1]:979. Cdk1 associates with M-cyclin as the levels of M-cyclin gradually rise. The resulting M-CDK complex is phosphorylated on an activating site by the CDK-activating kinase (CAK) and on a pair of inhibitory sites by the Wee1 kinase. The resulting inactive M-CDK complex is then activated at the end of G_2 by the phosphatase Cdc25. Cdc25 is further stimulated by active M-CDK, resulting in positive feedback.

G_2 and early mitosis, the M-CDK is activated by Cdc25, which removes the inhibitory phosphates restraining M-CDK. At the same time, Weel is suppressed, further ensuring the increase of M-CDK activity.

Interestingly, Cdc25 can also be activated, at least in part, by its target, M-CDK. M-CDK may also inhibit the inhibitory kinase Weel. The ability of M-CDK to activate its own activator (Cdc25) and inhibit its own inhibitor (Weel) suggests that M-CDK activation in mitosis involves positive feedback loops. According to this model, the partial activation of Cdc25 leads to the partial activation of a subpopulation of M-CDK complexes, which then phosphorylate Cdc25 and Weel molecules. This leads to more M-CDK activation, and so on. Such a mechanism would quickly promote the activation of all M-CDK complexes in the cell [1]:978. That is, in mitosis, M-CDK activates Cdc25 and inhibits Weel, involving positive feedback loops.

2.4.4 The APC/C Triggering Sister-Chromatid Separation and the Completion of Mitosis

In this subsection we describe the segregation of sister chromatids. After M-CDK has triggered the complex processes leading to metaphase, the cell cycle reaches its peak with the separation of the sister chromatids at the metaphase-to-anaphase transition. Although M-CDK activity sets the stage for this event, the anaphase-promoting complex (APC/C) throws the switch that initiates sister-chromatid separation by ubiquitylating several mitotic regulatory proteins and thereby triggering their destruction. That is, APC/C switches the initiation of sister-chromatid separation.

During metaphase, cohesins holding the sister chromatids together resist the poleward forces that pull the sister chromatids apart. Anaphase begins with the sudden

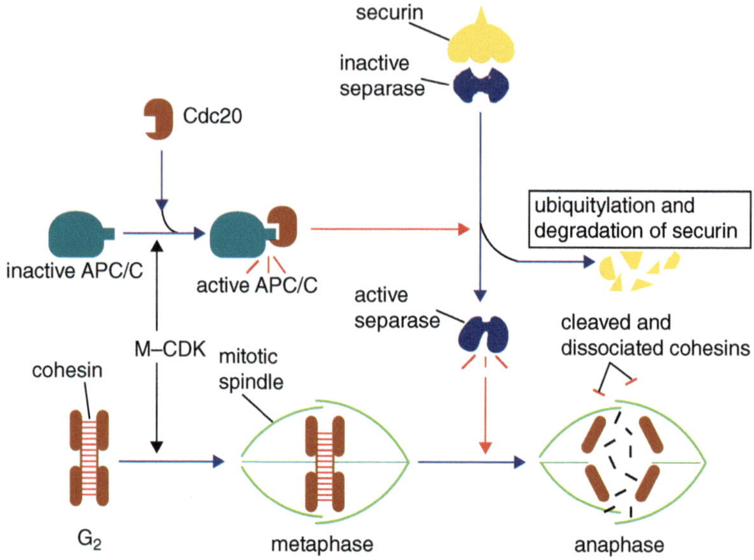

FIGURE 2.11 The initiation of sister-chromatid separation by the APC/C [1]:993. The activation of APC/C by Cdc20 leads to the ubiquitylation and destruction of securin, which normally holds separase in an inactive state. The destruction of securin allows separase to cleave Scc1, a subunit of the cohesin complex holding the sister chromatids together. The pulling forces of the mitotic spindle then pull the sister chromatids apart. Cohesin consists of four subsets. Two subsets, Smc1 and Smc3, are coiled coil proteins with an ATPase domain at one end; two additional subsets, Scc1 and Scc3, connect the ATPase head domains, forming a ring structure that may encircle the sister chromatids.

loss of sister-chromatid cohesion, which allows the sisters to separate and move to opposite poles of the spindle. The APC/C initiates the process by targeting the inhibitory protein securin for destruction. Before anaphase, securin binds to and inhibits the activity of a protease called separase. The destruction of securin at the end of metaphase releases separase, which is then free to cleave one of the subunits of cohesin. The cohesins fall away, and the sister chromatids separate (Figure 2.11) [1]:993. That is, during metaphase, cohesins holds the sister chromatids together. At the end of metaphase, APC/C initiates destruction of securin, releasing separase, which cleaves one of the subunits of cohesin. The cohesins falling away leads the sister chromatids to separate in anaphase.

In addition to securin, the APC/C also targets the S- and M-cyclins for destruction, leading to the loss of most CDK activity in anaphase. CDK inactivation allows phosphatases to dephosphorylate the many CDK target substrates in the cell, as required for the completion of mitosis and cytokinesis.

If the APC/C triggers anaphase, what activates the APC/C? The answer is only partly known. APC/C activation requires binding to the protein Cdc20. At least two processes regulate Cdc20 and its association with APC/C. Firstly, Cdc20 synthesis increases as the cell approaches mitosis, owing to an increase in the transcription of its gene. Secondly, phosphorylation of the APC/C helps Cdc20 bind to the APC/C, thereby helping to create an active complex. Among the kinases, it is M-CDK that phosphorylates and thus activates the APC/C. Thus, M-CDK not only triggers the early mitotic events leading to metaphase, but it also sets the stage for progression into anaphase [1]:993. That is, since M-CDK activates the APC/C, M-CDK not only triggers the early mitotic events leading to metaphase, but it also contributes to progression into anaphase as illustrated in Figure 2.11.

2.5 SEGREGATION AND CELL DIVISION

2.5.1 Segregated Chromosomes Packaged in Daughter Nuclei at Telophase

In this subsection, we describe the disassembly of the mitotic spindle and re-formation of nuclear envelope at telophase (the final stage of mitosis). By the end of anaphase, the daughter chromosomes have segregated into two equal groups at opposite ends of the cell. In telophase, the final stage of mitosis, the two sets of chromosomes are packaged into a pair of daughter nuclei.

The first major event of telophase is the disassembly of the mitotic spindle, followed by the re-formation of the nuclear envelope. Initially, nuclear membrane fragments associate with the surface of individual chromosomes. These membrane fragments fuse to partly enclose clusters of chromosomes and then coalesce to reform the complete nuclear envelope. Nuclear pore complexes are incorporated into the envelope, the nuclear lamina re-forms, and the envelope once again becomes continuous with the endoplasmic reticulum. That is, the events in telophase (the final stage of mitosis) are chromosome segregation and re-formation of the nuclear envelope. In telophase, the two sets of chromosomes are packaged into a pair of daughter nuclei.

Once the nuclear envelope has re-formed, the pore complexes pump in nuclear proteins, the nucleus expands, and the mitotic chromosomes are reorganized into their interphase state, allowing gene transcription to resume. A new nucleus has been created, and mitosis is complete. All that remains is for the cell to complete its division into two [1]:995.

2.5.2 Control of Cell Division and Cell Growth

In this subsection, we describe the control factor for size difference: cell growth, cell division, and cell survival. A fertilized mouse egg and a fertilized human egg are similar in size. However, they produce animals of very different sizes. What factors in the control of cell behavior in humans and mice are responsible for these size differences? The same fundamental question can be asked for each organ and tissue in an animal's body. What factors determine the length of an elephant's trunk or the size of its brain or its liver? It is possible to say what the ingredients of an answer to these questions must be.

The size of an organ or organism depends on its total cell mass, which depends on both the total number of cells and their size. The cell number, in turn, depends on the amounts of cell division and cell death.

Organ and body size are therefore determined by three fundamental processes: cell growth, cell division, and cell survival. Each is tightly regulated – both by intracellular programs and by extracellular signal molecules, which control these programs.

The extracellular signal molecules that regulate cell growth, division, and survival are generally soluble secreted proteins, proteins bound to the surface of cells, or components of the extracellular matrix. They can be divided operationally into three major classes:

1. Mitogens, which stimulate cell division, primarily by triggering a wave of G_1/S-CDK activity that relieves intracellular negative controls that otherwise block progress through the cell cycle.
2. Growth factors, which stimulate cell growth (an increase in cell mass) by promoting the synthesis of proteins and other macromolecules and by inhibiting their degradation.
3. Survival factors, which promote cell survival by suppressing the form of programmed cell death known as apoptosis [1]:1010. That is, mitogen is one of the extracellular signal molecules that regulate cell growth, division, and survival.

2.5.3 Mitogen Stimulating Cell Division

In this subsection, we describe that the cells of a multicellular organism divide only when they receive the stimulatory extracellular signal, mitogen. Unicellular organisms tend to grow and divide as fast as they can, and their rate of proliferation depends largely on the availability of nutrients in the environment. However, the cells of a multicellular organism divide only when the organism needs more cells. Thus, for an animal cell to proliferate, it must receive stimulatory extracellular signals, in the form of mitogens from other cells, usually its neighbors. Mitogens overcome intracellular breaking mechanisms that block progress through the cell cycle.

For the vast majority of animal cells, mitogens control the rate of cell division by acting in the G_1 phase of the cell cycle. Multiple mechanics act during G_1 to suppress CDK activity. Mitogens release these breaks on CDK activity, thereby allowing entry into a new cell cycle.

Mitogens interact with cell-surface receptors to trigger multiple intracellular signaling pathways. One major pathway acts through the monomeric GTPase Ras, which leads to the activation of a mitogen-activated protein kinase (MAP kinase) cascade. This leads to an increase in the production of transcription regulatory proteins, including MyC. MyC is thought to promote cell cycle entry by several mechanisms, one of which is to increase the expression of genes encoding G_1 cyclins (D-cyclins), thereby increasing G_1-CDK (cyclin D-CDK4) activity. MyC also has a major role in stimulating the transcription of genes that increase cell growth.

The key function of G_1-CDK complexes in animal cells is to activate a group of gene regulatory factors called the E2F proteins, which bind to specific DNA sequences in the promoters of a wide variety of genes that encode proteins required for S-phase entry, including G_1/S-cyclins, S-cyclins, and proteins involved in DNA synthesis and chromosome duplication. In the absence of mitogenic stimulation, E2F-dependent gene expression is inhibited by an interaction between E2F and members of the retinoblastoma protein (Rb) family. When cells are stimulated to divide by mitogens, active G_1-CDK accumulates and phosphorylates Rb family members, reducing their binding to E2F. The liberated E2F proteins then activate expression of their target genes (Figure 2.12).

This transcriptional control system, like so many other control systems that regulate the cell cycle, includes feedback loops that ensure that entry into the cell cycle is complete and irreversible. The liberated E2F proteins, for example, increase the transcription of their own genes. In addition, E2F-dependent transcription of G_1/S-cyclin (cyclin E) and S-cyclin (cyclin A) genes leads to increased G_1/S-CDK and

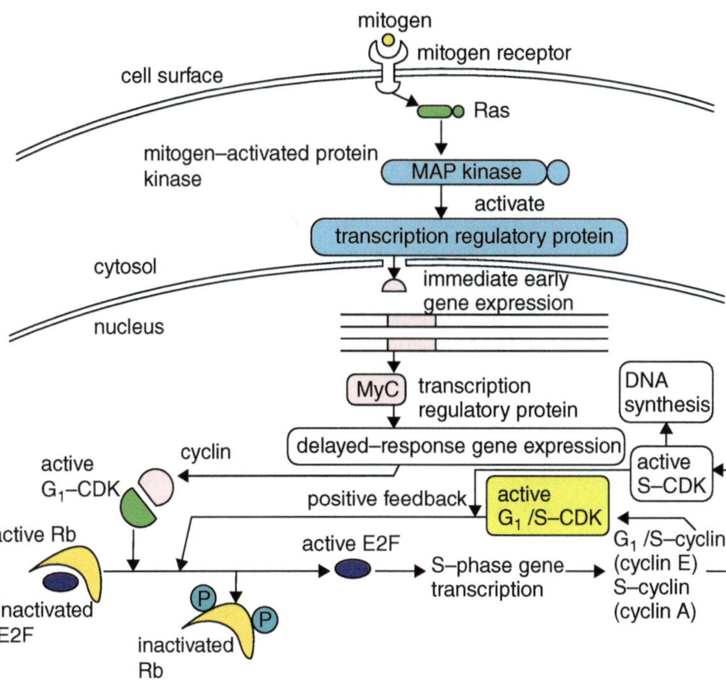

FIGURE 2.12 Mitogen stimulation of a cell cycle [1]:1013. Mitogens bind to cell-surface receptors to initiate intracellular signaling pathways. One of the major pathways involves activation of the small GTPase Ras, which activates a MAP kinase cascade, leading to increased expression of numerous immediate early genes, including the gene encoding the transcription regulatory protein MyC. MyC increases the expression of many delayed-response genes, including some that lead to increased G_1-CDK activity (cyclin D-CDK4), which triggers the phosphorylation of members of the Rb family of proteins. This inactivates the Rb proteins, freeing the gene regulatory protein E2F to activate the transcription of G_1/S genes, including the genes for a G_1/S-cyclin (cyclin E) and S-cyclin (cyclin A). The resulting G_1/S-CDK and S-CDK activities further enhance Rb protein phosphorylation, forming a positive feedback loop. E2F proteins also stimulate the transcription of their own genes, forming another positive feedback loop.

S-CDK activates, which in turn increase Rb protein phosphorylation and promote further E2F release (Figure 2.12), resulting in a positive feedback loop, leading to DNA synthesis [1]:1011.

2.6 UNFAVORABLE SITUATIONS IN THE CELL CYCLE

2.6.1 DNA Damage

In this subsection, we describe that unless the DNA damage is repaired, the DNA damage response can lead to either cell cycle arrest or cell death (apoptosis).

Progression through the cell cycle, and thus the rate of cell proliferation, is controlled not only by extracellular mitogens but also by other extracellular and intracellular signals. One of the most important influences is DNA damage, which can occur as a result of spontaneous chemical reactions in DNA, errors in DNA replication, or exposure to radiation or certain chemicals. It is essential that damaged chromosomes are repaired before attempting to duplicate or segregate them. The cell cycle control system can readily detect DNA damage and arrest the cycle at either of two transitions: one at Start, which prevents entry into the cell cycle and into the S phase, and one at the G_2/M transition, which prevents entry into mitosis.

DNA damage initiates a signaling pathway by activating one of a pair of related protein kinases called ATM (E2.6) and ATR, which associate with the site of damage and phosphorylate various target proteins, including two other protein kinases called Chk1 and Chk2 (E2.7). These various kinases phosphorylate other target proteins that lead to cell cycle arrest. A major target is the gene regulatory protein p53, which stimulates transcription of the gene encoding p21, a CKI protein (CDK inhibitor protein); p21 binds to G_1/S-CDK and S-CDK complexes and inhibits their activities, thereby helping to block entry into the cell cycle (Figure 2.13).

DNA damage activates p53 by an indirect mechanism. In undamaged cells, p53 is highly unstable and is present at very low concentrations. This is largely because it interacts with another protein, Mdm2, which acts as a ubiquitin ligase that targets p53 for destruction by proteasomes. Phosphorylation of p53 after DNA damage reduces its binding to Mdm2. This decreases p53 degradation, which results in a marked increase in p53 concentration in the cell. In addition, the decreased binding to Mdm2 enhances the ability of p53 to stimulate gene transcription (Figure 2.13).

Organisms take the following procedure if DNA damage is so severe that repair is not possible. In multicellular organisms, the health of the organism takes precedence over the life of an individual cell. Cells that divide with severe DNA damage threaten the life of the organism, since genetic damage can often lead to cancer and other diseases. Thus, animal cells with severe DNA damage do not attempt to continue division, but instead commit suicide by undergoing apoptosis. Thus, unless the DNA damage is repaired, the DNA damage response can lead to either cell cycle arrest or cell death. DNA damage-induced apoptosis often depends on the activation of p53. Indeed, it is this apoptosis-promoting function of p53 that is apparently most important in protecting us against cancer [1]:1014. That is, unless the DNA damage is repaired, the DNA damage leads to either cell cycle arrest or cell death (apoptosis) by the apoptosis-promoting function of p53.

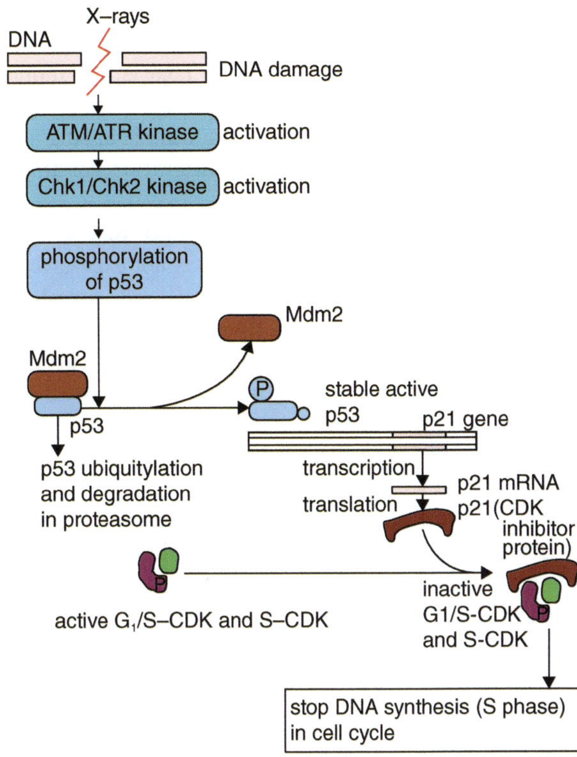

FIGURE 2.13 DNA damage arresting the cell cycle [1]:1015. When DNA is damaged, various protein kinases are recruited to the site of damage and initiate a signaling pathway that causes cell cycle arrest. The first kinase at the damage site is either ATM (E2.6) or ATR, depending on the type of damage. Additional protein kinases, called Chk1 and Chk2 (E2.7) are then recruited and activated, resulting in the phosphorylation of the transcription regulatory protein p53. Mdm2 normally binds to p53 and promotes its ubiquitylation and destruction in proteasomes. Phosphorylation of p53 blocks its binding to Mdm2, as a result. p53 accumulates to a high level and stimulates transcription of numerous genes, including the gene that encodes the CKI (CDK inhibitor protein) protein p21. The p21 binds and inactivates G_1/S-CDK and S-CDK complexes, arresting the cell in G_1. In some cases, DNA damage also induces either the phosphorylation of Mdm2 or a decrease in Mdm2 production, which causes a further increase in p53 (not shown).

2.6.2 Excessive Stimulation of Mitogens

In this subsection, we describe that the cell undergoes either permanent cell cycle arrest or apoptosis when there is excessive Ras or MyC production.

Many of the components of mitogenic signaling pathways are encoded by genes that were originally identified as cancer-promoting genes, because mutations in them contribute to the development of cancer. The mutation of a single amino acid in the small GTPase Ras, for example, causes the protein to become permanently overactive, leading to constant stimulation of Ras-dependent signaling pathways, even in the absence of mitogenic stimulation. Similarly, mutations that cause an overexpression of MyC stimulate excessive cell growth and proliferation and thereby promote the development of cancer. That is, the mutation in the Ras causes the protein to become permanently overactive, leading to constant stimulation of Ras-dependent signaling pathways. The mutation causing an overexpression of MyC stimulates excessive cell growth and proliferation, leading to the development of cancer.

Surprisingly, however, when a hyperactivated form of Ras or MyC is experimentally overproduced in most normal cells, the result is not excessive proliferation but the opposite: the cell undergoes either permanent cell cycle arrest or apoptosis. The normal cell seems able to detect abnormal mitogenic stimulation, and it responds by preventing further division. Such responses help prevent the survival and proliferation of cells with various cancer-promoting mutations. That is, the normal cell can detect abnormal mitogenic stimulation such as overactive Ras and overexpression of MyC, and it responds by preventing further division, leading to permanent cell cycle arrest or apoptosis.

Although it is not known how a cell detects excessive mitogenic stimulation, such stimulation often leads to the production of a cell cycle inhibitor protein called Arf, which binds and inhibits Mdm2. Mdm2 normally promotes p53 degradation. Activation of Arf therefore causes p53 levels to increase, inducing either cell cycle arrest or apoptosis (Figure 2.14). That is, activation of Arf causing increase of p53 induces either cell cycle arrest or apoptosis.

How do cancer cells ever arise even if these mechanisms block the division or survival of mutant cells with overactive proliferation signals? The answer is that the protective system is often inactivated in cancer cells by mutations in the genes that encode essential components of the blocking mechanisms, such as the inhibitor protein Arf or p53 or the proteins that help activate them [1]:1016.

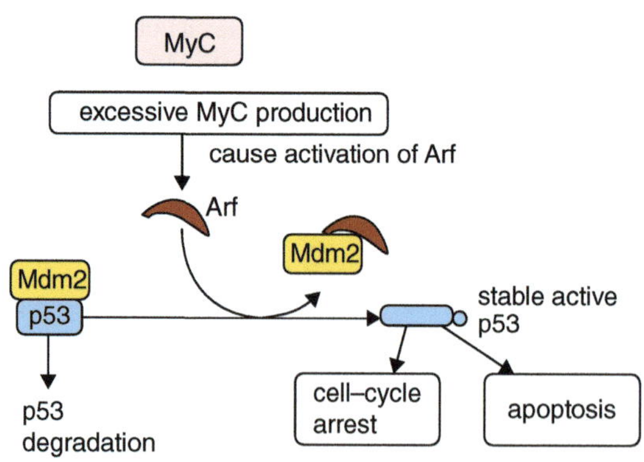

FIGURE 2.14 Cell cycle arrest or apoptosis induced by excessive stimulation of mitogenic pathways [1]:1017. Abnormally high levels of MyC cause the activation of Arf, which binds and inhibits Mdm2 and thereby increases p53 levels. Depending on the cell type and extracellular conditions, p53 then causes either cell cycle arrest or apoptosis.

2.7 CELL PROLIFERATION COORDINATING GROWTH AND DIVISION

2.7.1 Cell Proliferation Accompanied by Cell Growth

In this subsection, we describe that cell growth accompanies cell division, and in animals, cell growth and cell proliferation depend on extracellular signal molecules called growth factors and mitogens, respectively. Cells proliferating without growing get progressively smaller and there is no net increase in total cell mass. In most proliferating cell populations, therefore, cell growth accompanies cell division. In single-celled organisms such as yeasts, both cell growth and cell division require only nutrients. In animals, by contrast, both cell growth and cell proliferation depend on extracellular signal molecules, produced by other cells, which are called growth factors and mitogens, respectively.

Like mitogens, the extracellular growth factors that stimulate animal cell growth bind to receptors on the cell surface and activate intracellular signaling pathways. These pathways stimulate the accumulation of proteins and other macromolecules, and they do so by both increasing their rate of synthesis and decreasing their rate of degradation. They also trigger increased uptake of nutrients and production of the ATP required to fuel the increased protein synthesis. One of the most important intracellular signaling pathways activated by growth factor receptors involves the enzyme phosphoinositide 3-kinase (PI3-kinase), which adds a phosphate from ATP to the 3' position of inositol phospholipids in the plasma membrane. The activation of PI3-kinase leads to the activation of a kinase called TOR (target of rapamycin) (see Figure 2.15), which lies at the heart of cell growth regulatory pathways in all eukaryotes. TOR activates many targets in the cell that stimulate metabolic processes, including protein synthesis. One target is a protein kinase called S6-kinase (S6K), which phosphorylates ribosomal protein S6, increasing the ability of ribosomes to translate a subset of mRNAs that mostly encode ribosomal components. TOR also indirectly activates a translation initiation factor called eIF4E and directly activates transcription regulators that promote the increased expression of genes encoding ribosomal subunits (Fig. 2.15) [1]:1017. That is, like mitogen, the extracellular growth factor binds to receptor on the cell surface and activate intracellular signaling pathways, stimulating the accumulation of proteins and other macromolecules, and increasing their rate of synthesis. One of the most important intracellular signaling pathways involves the enzyme PI3-kinase, whose activation leads to the activation of kinase TOR. TOR activates S6K-kinase, increasing the ability of ribosomes. TOR also activates a translation initiation factor called eIF4E.

2.7.2 Proliferating Cells Coordinate Their Growth and Division

In this subsection, we describe that to maintain a constant size, proliferating cells must coordinate their growth with cell division, but animal cell growth and division are not always coordinated.

For proliferating cells to maintain a constant size, they must coordinate their growth with cell division to ensure that cell size doubles with each division: if cells

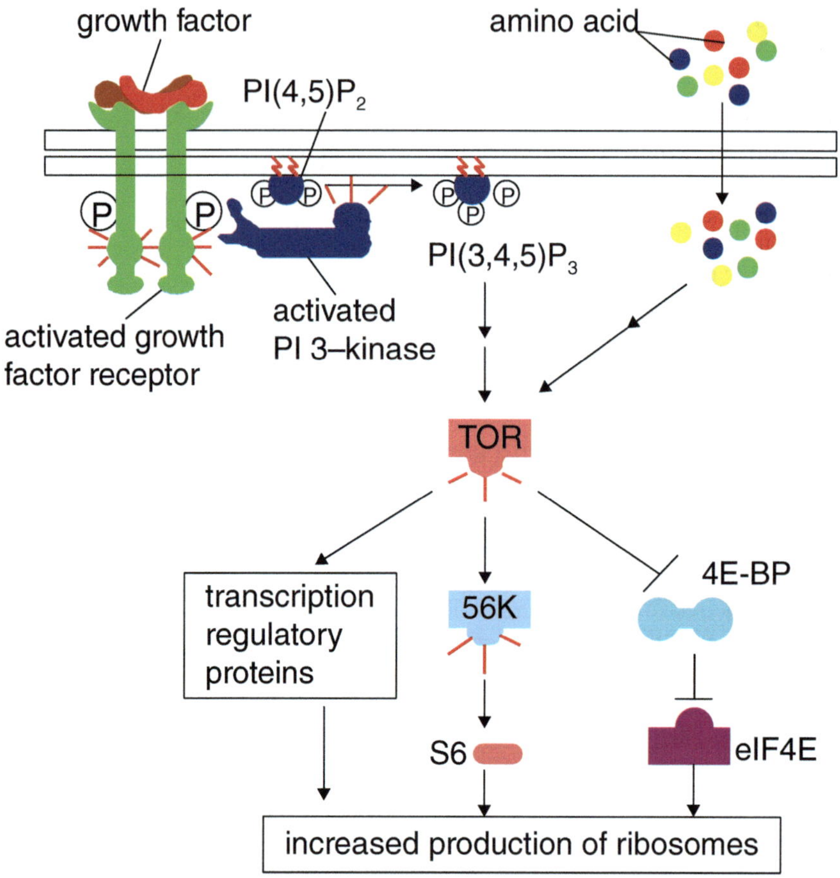

FIGURE 2.15 Stimulation of cell growth by extracellular growth factors and nutrients [1]:1017. The occupation of cell surface receptors by growth factors leads to the activation of PI3-kinase, which promotes protein synthesis through a complex signaling pathway such as PI(4,5)P_2 and PI(3,4,5)P_3 that leads to the activation of the protein kinase TOR; extracellular nutrients such as amino acids also help activate TOR. TOR phosphorylates multiple proteins to stimulate protein synthesis, as shown; it also inhibits protein degradation (not shown). Growth factors also stimulate increased production of the transcription regulatory protein MyC (not shown), which activate the transcription of various genes that promote cell metabolism and growth. 4E-BP is an inhibitor of the translation initiation factor eIF4E. PI3-kinase activated by active growth factor receptor, promotes TOR through the signaling pathway consisting of PI(4,5)P_2, phosphatidylinositol 4,5-bisphosphate and PI(3,4,5)P_3, phosphatidylinositol 3,4,5-triphosphate.

grow too slowly, they will get smaller with each division, and if they grow too fast, they will get larger with each division. It is not clear how cells achieve this coordination, but it is likely to involve multiple mechanisms that vary in different organisms and even in different cell types of the same organism (Figure 2.16).

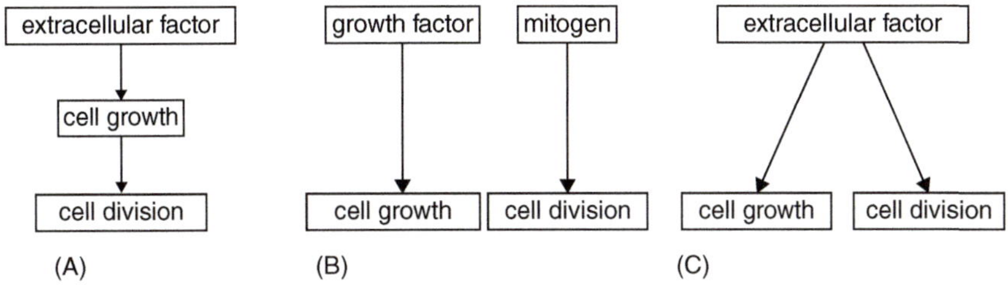

FIGURE 2.16 Potential mechanisms for coordinating cell growth and division [1]:1018. In proliferating cells, cell size is maintained by mechanisms that coordinate rates of cell division and cell growth. Numerous alternative coupling mechanisms are thought to exist, and different cell types appear to employ different combinations of these mechanisms. (A) In many cell types – particularly yeast – the rate of cell division is governed by the rate of cell growth, so that division occurs only when the growth rate achieves some minimal threshold; in yeast, it is mainly the level of extracellular nutrients that regulate the rate of cell growth and thereby the rate of cell division. (B) In some animal cell types, growth and division can each be controlled by separate extracellular factors (growth factors and mitogens, respectively) and cell size depends on the relative levels of two types of factors. (C) Some extracellular factors can stimulate both cell growth and cell division by simultaneously activating signaling pathways that promote growth and other pathways that promote cell cycle progression.

Animal cell growth and division are not always coordinated, however. In many cases, they are completely uncoupled to allow growth without division or division without growth. Muscle cells and nerve cells, for example, can grow dramatically after they have permanently withdrawn from the cell cycle. Similarly, the eggs of many animals grow to an extremely large size without dividing; after fertilization, however, this relationship is reversed, and many rounds of division occur without growth.

Compared to cell division, there has been surprisingly little study of how cell size is controlled in animals. As a result, it remains a mystery how cell size is determined and why different cell types in the same animals grow to be so different in size. One of the best-understood cases in mammals is the adult sympathetic neuron, which has permanently withdrawn from the cell cycle. Its size depends on the amount of nerve growth factor (NGF) secreted by the target cells it innervates; the greater the amount of NGF the neuron has access to, the larger it becomes. It seems likely that the genes a cell expresses set limits on the size it can be, while extracellular signal molecules and nutrients regulate the size within these limits. The challenge is to identify the relevant genes and signal molecules for each cell type [1]:1018.

E2.1 MITOGENS

One of the first mitogens to be identified was platelet-derived growth factor (PDGF), and it is typical of many others discovered since. The path to its isolation begins with the observation that fibroblasts in a culture dish proliferate when provided with serum but not when provided with plasma. Plasma is prepared by removing the cells from blood without allowing clotting to occur; serum is prepared by allowing blood to clot and taking the cell-free liquid that remains. When blood clots, platelets incorporated in the clot are stimulated to release the contents of their secretory vesicles. The super-ability of serum to support cell proliferation suggested that platelets contain one or more mitogens. This hypothesis was confirmed by showing that extracts of platelets could serve instead of serum to stimulate fibroblast proliferation. The crucial factor in the extracts was shown to be a protein, which was subsequently purified and named PDGF. In the body, PDGF liberated from blood clots helps stimulate cell division during wound healing.

PDGF is only one of over 50 animal proteins that are known to act as mitogens. PDGF, for example, can stimulate many types of cells to divide, including fibroblasts, smooth muscle cells, and neuroglial cells also called "glia" which is a Greek word meaning "glue." Neuroglia is seen in nervous tissues as cells surrounding the neurons. Many mitogens, including PDGF, also have other actions than the stimulation of cell division. They can stimulate cell growth, survival, differentiation, or migration, depending on the circumstances and the cell type [1]:1011.

E2.2 CYCLIN: CDK REGULATOR

Central components of the cell cycle control system are members of a family of protein kinases known as CDKs (cyclin-dependent kinases). The activities of these kinases rise and fall as the cell progresses through the cycle.

Cyclical changes in CDK activity are controlled by a complex array of enzymes and other proteins. The most important of these CDK regulators are proteins known as "cyclins." When cyclin forms a complex with CDK, the protein kinase is activated to trigger specific cell cycle events. Without cyclin, CDK is inactive [1]:968.

E2.3 PROTEIN DIGESTION BY PROTEASOME

Due to mutations or errors in transcription and translation, some proteins are never produced properly, and it is particularly important that the cell should destroy these potentially harmful proteins. The apparatus that deliberately destroys aberrant proteins is the proteasome. Figure E2.1 shows the process of destroying ubiquitin-marked protein by proteasome [1]. As shown in Figure 2.13, if p53 is not phosphorylated

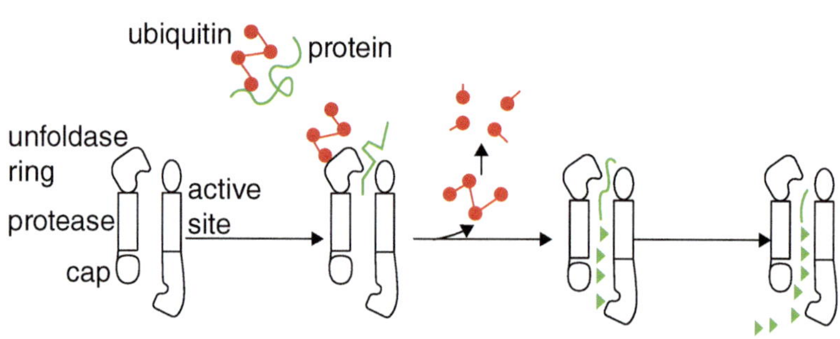

FIGURE E2.1 Protein digestion by proteasome [1]:358.

and p53 inactivated by Mdm2 is not necessary, ubiquitin-marked p53 is digested by proteasome.

E2.4 PHOSPHORYLATION

Without mitogenic stimulation, E2F-dependent gene expression is inhibited by an interaction between E2F and members of the Rb (retinoblastoma protein) family. When mitogens stimulate cells to divide, active G_1-CDK accumulates and phosphorylates Rb family members, reducing their binding to E2F. The liberated E2F proteins then activate expression of their target genes (Figure 2.12) [1]:1012.

Unless the DNA damage is repaired, the DNA damage response can lead to either cell cycle arrest or cell death. The DNA damage-induced apoptosis often depends on the activation of p53 by its phosphorylation. Indeed, it is this apoptosis-promoting function of p53 that is apparently important in protecting us against cancer (Figure 2.13) [1]:1015. Thus, phosphorylation of proteins concerning a cell cycle leads to controlling such proteins: active or inactive. For example, phosphorylation of Rb leads to inactive Rb protein, inducing active E2F protein promoting DNA synthesis. Another phosphorylation of p53 leads to inactive CDK, inducing cell death.

E2.5 APOPTOSIS

The growth, development, and maintenance of multicellular organisms depend not only on the production of cells but also on mechanisms to destroy them. The maintenance of tissue size, for example, requires that cells die at the same rate as they are produced. During development, carefully orchestrated patterns of cell death help determine the size and shape of limbs and other tissues. Cells also die when they become damaged or infected, removing threatening health of the organism. Cell death occurs by a programmed sequence of molecular events, systematically destroying itself. In most cases, the programmed cell death occurs by a process called "apoptosis," from the Greek word meaning "falling off," as leaves from a tree [1]:1021.

Cells that die as a result of acute injury swell and burst, spilling their contents all over their neighbors: a process called "necrosis." This eruption triggers a potentially damaging inflammatory response. By contrast, a cell that undergoes apoptosis dies neatly without damaging its neighbors. The molecular machinery responsible for apoptosis involves a family of proteases called "caspases." These enzymes are made as inactive precursors, called procaspases, which are activated in response to signals inducing apoptosis [2]:635.

E2.6 ATM PROTEIN FOR DNA REPAIR

The DNA damage results in an increased synthesis of some DNA repair enzymes. This response depends on special signaling proteins that sense the DNA damage and up-regulate the appropriate DNA repair enzymes. The ATM protein is a large kinase needed to generate the intracellular signals that sound the alarm in response to many types of spontaneous DNA damages. Humans born with defects in the

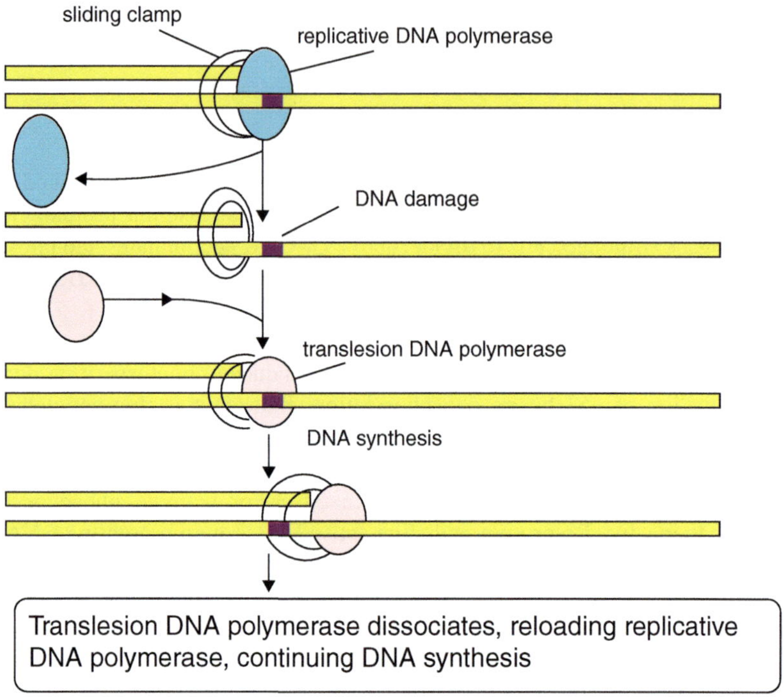

FIGURE E2.2 Translesion DNA polymerases can use damaged templates [1]:274.

gene encoding the ATM protein have the genetic disease ataxia telangiectasia (AT) [1]:1015, the symptoms of which include neurodegeneration, a predisposition to cancer, and genome instability. The mechanism concerning the ATM protein needed for the DNA repair mentioned earlier is revealed by the phenotype of humans with defects in the gene encoding the ATM protein [1]:276. That is, since humans with defects in the gene encoding the protein ATM have the genetic disease, the mechanism of the DNA repair is revealed by the phenotype of humans with defects in the gene encoding the ATM protein.

A conceptual model ([1]:274) of the DNA repair process is given as follows. When a replicative DNA polymerase encounters DNA damage, specialized enzymes covalently modify the sliding clamp which releases the replicative DNA polymerase, and together with damaged DNA, attracts a translesion DNA polymerase specific to the type of damage (Figure E2.2). The translesion DNA polymerase is responsible for most of the base-substitution and single-nucleotide deletion mutations that accumulate in genomes. Once the damaged DNA is bypassed, the covalent modification of the clamp is removed, and the translesion DNA polymerase dissociates. And then the replicative DNA polymerase is brought back into play, continuing DNA synthesis. That is, during the DNA repair process, at the DNA damage the sliding clamp releases the replicative DNA polymerase and attracts a translesion DNA polymerase to damaged DNA. The translesion DNA polymerase is responsible for the base-substitution and single-nucleotide deletion mutations. Once the damaged DNA is bypassed, the translesion DNA polymerase dissociates. Then the replicative DNA polymerase is brought back, continuing DNA synthesis.

E2.7 CHK1, CHK2, AND P53 FOR CYCLE ARREST

Unless the DNA damage is repaired, the DNA damage response can lead to either cell cycle arrest or cell death. The DNA damage-induced apoptosis often depends on the activation of p53, protecting us against cancer. Chk1 and Chk2 kinases phosphorylate the gene regulatory protein p53, which stimulates transcription of the gene encoding p21, a CKI (CDK inhibitor) protein; p21 binds to G_1/S-CDK and S-CDK complexes and inhibits their activities, helping to block entry into the cell cycle [1]:1015.

3 The Immune System

The binding of an antigen to BCRs (B-cell receptors) activates B-cells, secreting antibodies. Antibodies protect against pathogens by binding them and blocking their access to cells which may be infected and destroyed. This is known as neutralization. Epitope (antigenic determinant) buried within the antigen, displayed on the surface of the antigen-presenting cell (APC), is recognized by cytotoxic T-cells killing APC.

Immunology is the study of the body's defense against infection. We are always exposed to microorganisms, a number of which are the cause of infectious disease. How does the body protect against microorganisms? When infection arises, how does the body eliminate invaders, and cure disease? Why do we develop long-lasting immunity after overcoming an infectious disease we encountered only once? These are questions answered by immunology. Immunology is the study of understanding – at cellular and molecular levels – the body's defenses against infection.

3.1 HISTORICAL BACKGROUND

3.1.1 The Beginning of Immunology

As a science, the beginning of immunology is attributed to the work of Edward Jenner in the late eighteenth century. E. Jenner observed that the relatively mild disease of cowpox seemed to often confer protection against the fatal disease of smallpox. In 1796, he demonstrated that inoculation with cowpox protected the recipient against smallpox. His scientific proof relied on a careful plan of exposing inoculated individuals to infectious smallpox material for two months after inoculation. E. Jenner called this procedure "vaccination." This term is used to express inoculating healthy individuals with weakened strains of disease-causing agents in order to protect against disease.

E. Jenner's strategy of vaccination was extended by many great microbiologists' discoveries in the late nineteenth century. Robert Koch (E3.2) proved that infectious disease is caused by specific microorganisms. In the 1880s, Louis Pasteur devised a vaccine against cholera in chickens. He then developed a rabies vaccine; when he vaccinated a boy bitten by rabid dog, he proved the striking success.

These successes led to a search for vaccination's mechanism of protection and to the development of the science of immunology. In the early 1890s, Emil von Behring and Shibasaburo Kitasato (E3.3) discovered specific antitoxic activity that is contained in the serum of animals immune to diphtheria or tetanus. The specific antitoxic activity was capable of conferring short-lived protection against the effects of diphtheria or tetanus toxins in people. The activity was later determined to be due to proteins called antibodies, which bind specifically to the toxins and neutralize their activity. The crucial role of these antibodies in immunity was reinforced by Jules

DOI: 10.1201/9781003780663-3

Bordet's discovery in 1899 of complement, a component of serum that acts in conjunction with antibodies to destroy pathogenic microorganisms.

Adaptive immunity is a specific response against infection by potential pathogens, such as the production of antibodies against a particular pathogen, because it develops during the lifetime of an individual as an adaptation to infection with that pathogen. Adaptive immunity is distinguished from innate immunity, which was already known at the time that von Behring developed serum therapy for diphtheria chiefly through the work of the great Russian immunologist Elie Metchnikoff, who discovered that many microorganisms could be engulfed and digested by phagocytic cells. The phagocytic cells provide defenses against nonspecific infection. E. Metchnikoff called the phagocytic cells "macrophages," (Greek "phagein," meaning "to eat"). The macrophages are always naturally present and ready to act, playing an important role in the early immune response. On the other hand, adaptive immunity requires time to develop but is highly specific.

It was soon clear that specific antibodies could be induced against a vast range of substances, called antigens because they could stimulate antibody generation. Paul Ehrlich advanced the development of an antiserum as a treatment for diphtheria and developed methods to standardize therapeutic serums. Today the term antigen refers to any substance recognized by the adaptive immune system. Typically antigens are common proteins, glycoproteins, and polysaccharides of pathogens, but they can include a much wider range of chemical structures; for example, metals such as nickel, drugs such as penicillin, and organic chemicals such as urushiol (a mix of pentadecylcatechols [3]:2) in the leaves of poison ivy. Metchnikoff and Ehrlich were awarded the Nobel Prize for their work on immunity.

3.1.2 Microorganisms

We recognize four categories of disease-causing microorganisms or pathogens: viruses, bacteria, fungi, and parasites. The smallest are viruses which range from five to a few hundred nanometers in size, and the largest pathogens are parasites. Viruses are obligate intracellular pathogens and can directly kill cells by inducing lysis during their replication. Somewhat larger intracellular bacteria and microbacteria directly kill cells or damage cells by producing toxins. Many single-celled intracellular parasites, such as members of the *Plasmodium* genus that cause malaria, also directly kill infected cells. Pathogenic bacteria and fungi growing in extracellular spaces can induce shock and sepsis by releasing toxins into the blood or tissues. The largest pathogens – parasitic worms, or helminths – are too large to infect host cells. They can, however, injure tissue by forming cysts which induce damaging cellular responses in tissues into which the worms migrate.

Not all microbes are pathogens: there are microorganisms useful for human living (E3.1). Many tissues, especially mucous membranes such as oral mucosa, conjunctiva, and the lining of the gastrointestinal tract, are constantly colonized by microbial communities, called the microbiome, which consist of archaea, bacteria, and fungi that cause no damage to the host. They are also called commensal microorganisms, since they can have a symbiotic relationship with the host. Indeed, some commensal organisms perform important functions, as in the case of bacteria that aid in cellulose digestion in the stomachs of ruminants. Even enormous numbers of microbes in the intestinal microbiome normally cause no damage and are confined within the intestinal lumen by a protective layer of mucus, whereas pathogenic bacteria can penetrate this barrier, injure intestinal epithelial cells, and spread into the underlying tissues.

3.2 INNATE IMMUNITY

3.2.1 Innate Immunity Being the Initial Defense Against Infection

The human body is naturally endowed with innate immunity. The host cells adopt three strategies to deal with the threat posed by microbes: avoidance, resistance, and tolerance. The avoidance mechanism prevents exposure to microbes by a barrier of tissue such as skin. If the barrier cannot protect infection, the in established infection, the objective becomes resistance, reducing pathogens and eliminating them. In order to defend against the great variety of microbes, the immune system has many molecular and cellular functions called "mediator" or "effector" mechanisms suited to resist different categories of pathogens.

Tolerance involves a response enhancing tissue's capacity to resist damage induced by microbes [3]:5.

Anatomic and chemical barriers are the initial defenses against infection. The skin and mucosal surfaces represent a kind of avoidance strategy that prevents exposure of internal tissues to microbes. At most anatomic barriers, additional resistance mechanisms further strengthen host defenses. For example, mucosal surfaces produce a variety of antimicrobial proteins that act as natural antibiotics to prevent microbes from entering the body.

If these barriers are breached or evaded, other components of the innate immune system can immediately come into play. The complement system discovered by Jules Bordet is a group of around 30 different plasma proteins that act together and are one of the most important effector mechanisms in serum and interstitial tissues. The complement system not only acts in conjunction with antibodies, but can also target foreign organisms in the absence of a specific antibody; thus it contributes to both innate and adaptive responses.

Pathogens breaching the host's anatomic barrier encounter the cellular defenses of innate immunity. Cellular immune responses are initiated when sensor cells detect inflammatory inducers. Inflammatory inducers are chemical structures that indicate the presence of invading microbes or the cellular damage produced by them. Sensor cells include many cell types that detect inflammatory mediators through expression of many innate recognition receptors, which are encoded by a relatively small number of genes that remain constant over an individual's lifetime. Inflammatory inducers that trigger these receptors include molecular components unique to bacteria or viruses, such as bacterial lipopolysaccharides, or molecules such as ATP, which is not normally in the extracellular space. Triggering these receptors can activate innate immune cells to produce various mediators that either act directly to destroy invading microbes, or act on other cells to propagate the immune response. For example, macrophages can ingest microbes and produce toxic chemical mediators, such as degradative enzymes or reactive oxygen intermediates, to kill them. Dendritic cells may produce cytokine mediators, including many cytokines that activate target tissues, such as epithelial or other immune cells, to resist or kill invading microbes more efficiently. That is, when pathogens breach the host's anatomic barrier, cellular immune responses are initiated as follows. Sensor cells such as macrophages, neutrophils, and dendritic cells detect inflammatory inducers which have a chemical structure indicating the presence of invading microbes or the cellular damage produced by them, and which include molecular components unique to bacteria or viruses. The triggered sensor cells produce mediators such as cytokines and chemokines. For example, dendritic cells produce cytokine mediators activating target tissues to resist or kill invading microbes.

The common myeloid precursor (CMP) is the precursor of macrophages, granulocytes (the collective term for the white blood cells called neutrophils, eosinophils, and basophils), mast cells, and dendritic cells of the innate immune system (Figure E3.1). Macrophages, granulocytes, and dendritic cells make up the three types of phagocytes in the immune system [3]:7–8. The CMP also generates megakaryocytes and red blood cells.

3.2.2 Functions of Interferons

Viral infection in host cells induces the production of interferons, originally named because of their ability to interfere with viral replication in uninfected cells. Interferons help to defend against viral infection in several ways. The interferons IFN-α and IFN-β have three major functions.

Firstly, they induce resistance to viral replication in uninfected cells by activating genes that cause the destruction of mRNA and inhibit the translation of viral proteins. IFN-β is particularly important because it induces cells to make IFN-α, thus amplifying the interferon response. Interferons act to induce a state of resistance to viral replication in all cells. IFN-α and IFN-β bind to a common cell-surface receptor, known as the interferon-α receptor (IFNAR), which uses the JAK (Janus kinase) and STAT (signal transducers and activators of transcription) pathway. IFNAR uses the kinases Tyk2 and Jak1 to activate the factors STAT1 and STAT2, which translocate into the nucleus to interact with IRF9 and form a complex called ISGF3 that binds to the promoters of many interferon-stimulated genes (ISGs). One ISG encodes the enzyme oligoadenylate synthetase, which polymerizes ATP into 2'5'-linked oligomers. These oligomers activate an endoribonuclease that then degrades viral RNA. A second protein induced by IFN-α and INF-β is a dsRNA-dependent protein kinase called PKR, suppressing protein translation and contributing to the inhibition of viral replication. IFIT (IFN-induced protein with tetratricoid repeats) functions in suppressing translation of viral RNA [3]:110, 122. That is, the first function of the interferons is to induce resistance to viral replication in uninfected cells. IFIT protein functions in suppressing translation of viral RNA.

Secondly, they can induce MHC class I expression (section 3.8) in most cell types in the body, making them more susceptible to being killed by CD8 cytotoxic T-cells (section 3.9). That is, the second function is to induce MHC class I expression and antigen presentation.

Thirdly, they activate NK (natural killer) cells (section 3.5), which then selectively kill virus-infected cells [3]:122.

3.2.3 Cells in the Innate Immune Response

The cellular immune response initiates when sensor cells detect inflammatory inducers. Active sensor cells directly act to destroy invading microbes, or produce mediators acting on other cells to propagate the immune response. Macrophages, which are one of the sensor cells, ingest microbes and produce toxic chemical mediators such as degradative enzymes to kill microbes.

3.2.4 Phagocytic Cell

If a microorganism crosses an epithelial barrier and begins to replicate in the tissues of the host, it is immediately recognized by resident phagocytic cells. The main classes of phagocytic cells in the innate immune system are macrophages, monocytes, granulocytes, and dendritic cells. Macrophages are the major phagocyte population resident in most normal tissues at homeostasis. They can arise either from progenitor cells that enter the tissues during embryonic development, and then self-renew at steady state during life, or from circulating monocytes. During infection or inflammation, macrophages can also arise from monocytes that leave from the circulation to enter into tissues. Monocytes develop in the bone marrow and circulate in the blood. In humans, 90% of circulating monocytes are the classical monocyte that expresses CD14, a co-receptor for a PRR (pattern recognition receptor), and function during infection by entering tissues and differentiating into activated inflammatory monocytes or macrophages. PRRs function as sensors for pathogen-associated molecular patterns (PAMPs). Cytokine and chemokine production by macrophages is the result of signaling by these PRRs that is induced by a wide variety of pathogen components.

The second major family of phagocytes comprises the granulocytes, which include neutrophils, eosinophils, and basophils. Of these, neutrophils have the greatest phagocytic activity and are the cells most immediately involved in innate immunity against infection agents. Macrophages and granulocytes have an important role in innate immunity because they can recognize, ingest, and destroy many pathogens without the aid of an adaptive immune response.

The third class of phagocytes in the immune system is the dendritic cells that reside in lymphoid organs and in peripheral tissues. They arise from progenitors within the bone marrow that primarily branch from cells of myeloid potential, and they migrate via the blood to tissues throughout the body and to peripheral lymphoid organs. Dendritic cells ingest and break down microbes, but unlike macrophages and neutrophils, their primary role in immune defense is not the frontline, large-scale direct killing of microbes. A major role of dendritic cells is to process ingested microbes in order to generate peptide antigens that can activate T-cells and induce an adaptive immune response. Dendritic cells are thus considered to act as a bridge between innate and adaptive immune responses.

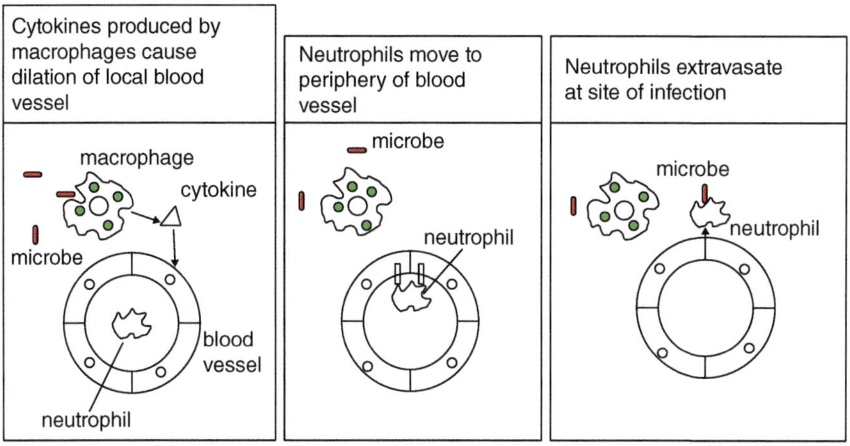

FIGURE 3.1 Infection stimulates macrophages to release cytokines that lead to movement of neutrophils, extravasating from blood vessels [3]:86.

Macrophages and dendritic cells (E3.5) produce cytokine mediators. Cytokine (E3.6) is material propagating information, and activates other immune cells to effectively resist or kill invading microbes [3]:6. Neutrophils are the most numerous and most important cells in the innate immune response, and take up a variety of microorganisms, effectively destroying them in intracellular vesicles by using degradative enzymes (Figure 3.1) [3]:86, 116. NK cells have TRAIL (tumor necrosis factor-related apoptosis-inducing ligand) on its surface, leading to apoptosis of the target cell (as shown in Figure 3.3 in the following section).

3.2.5 NK Cells

A defining feature of adaptive immunity is the clonal expression of antigen receptors, that provide the extraordinarily diverse specificities of T and B lymphocytes. However, for several decades, immunologists have recognized cells that have lymphoid characteristics but lack specific antigen receptors. These are now called innate lymphoid cells (ILCs) and include NK cells. ILCs develop in the bone marrow from the same common lymphocyte progenitor (CLP) that gives rise to B- and T-cells. ILCs are identified by the absence of T- and B-cell antigen receptor and co-receptor complexes, but they express the receptor for IL–7. They migrate from the bone marrow and populate lymphoid tissues and peripheral organs, notably the dermis, liver, small intestine, and lungs. NK cells are larger than T- and B-cells, have distinctive cytoplasmic granules containing cytotoxic proteins, and are functionally identified by their ability to kill certain tumor cell lines without the need for specific immunization. NK cells kill cells by releasing their cytotoxic granules, which are similar to those of T-cells and have the same effect. However, unlike T-cells, killing by NK cells is triggered by germline-encoded receptors that recognize molecules on the surface of infected or malignantly transformed cells [3]:11, 124. That is, the CLP in the

Phase of the immune response		
Response	Typical time after infection to start of response	Duration of response
Innate immune response	minutes	days
Adaptive immune response	a few days	weeks
Immunological memory	days to weeks	can be lifelong

FIGURE 3.2 Phases of the immune response [3]:6.

bone marrow gives rise both to antigen-specific lymphocytes of the adaptive immune system and to several innate lineages that lack antigen-specific receptors. Unlike T-cells, NK cells' killing is triggered by germline-encoded receptors recognizing molecules on the surface of infected cells.

Natural killer cells in the innate immune system were not discovered until the 1970s. NK cells lack the antigen-specific receptors, but express various families of innate receptors. NK cells play an important role in the early innate immune response (Figure 3.2) to viral infections [3]:11. NK cells are activated by cytokine IL-12 secreted by macrophages and dendritic cells during infection by many types of pathogens [3]:111, 125. That is, NK cells lack the antigen-specific receptors, but express various families of innate receptors.

In the absence of MHC class I, NK cells destroy cells by releasing cytotoxic granules whose contents contain granzymes and the pore-forming protein perforin, to the target cell membrane, penetrating into a target cell and inducing programmed cell death (Figure 3.3 top and bottom left). Killing by NK cells is triggered by activating receptors recognizing the ligands for the receptors, expressed in response to cellular stress such as infection, tumor, and injury [3]:130. A second pathway in killing by NK cells involves the TNF (tumor necrosis factor) family member known as TRAIL that interacts with two TNF superfamily death receptors, DR4 and DR5. When NK cells recognize a target cell, TRAIL activates DR4 and DR5 which signal through FADD (Fas-associated via death domain) [3]:471, an adaptor that activates pro-caspase8, leading to introduction of apoptosis of the target cell (Figure 3.3 bottom right) [3]:127. Fas is a TNF receptor [3]:796. That is, in the absence of MHC class I, NK cells destroy target cells by releasing cytotoxic granules. The second pathway in killing by NK cells is TRAIL that interacts with DR4 and DR5 which signal through FADD, an adaptor that activates pro-caspase8, leading to apoptosis of the target cell.

On the other hand, in the presence of MHC class I, inhibitory receptors on NK cells that recognize MHC class I molecules, function to prevent NK cells from killing normal host cells.

Signaling by the inhibitory NK receptors suppresses the killing activity and cytokine production of NK cells. This means that NK cells will not kill healthy, genetically identical cells with normal expression of MHC class I molecules, such as the other cells of the body. Virus-infected cells, however, can become susceptible to being killed by NK cells by a variety of mechanisms. Firstly, some viruses inhibit all

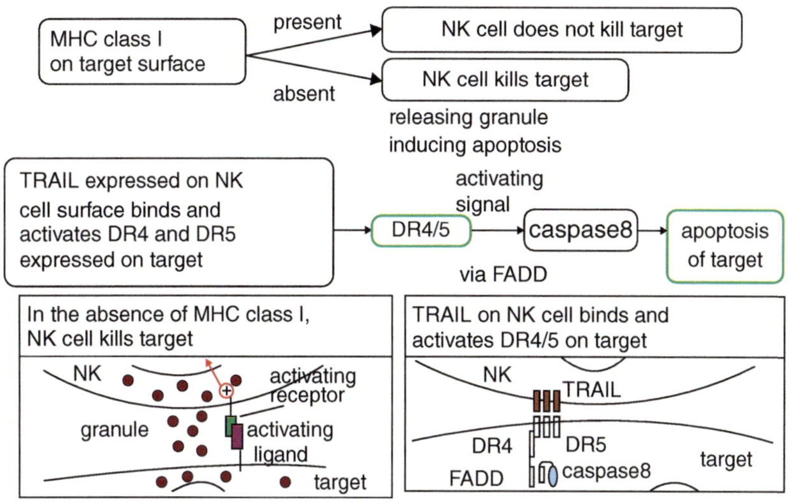

FIGURE 3.3 Killing target cells by NK cells [3]:127.

protein synthesis in their host cells, so that synthesis of MHC class I proteins would be blocked in infected cells, even while their production in uninfected cells is being stimulated by the actions of type I interferons. The reduced level of MHC class I expression in infected cells would make them correspondingly less able to inhibit NK cells through their MHC-specific receptors, and they would become more susceptible to being killed. Secondly, many viruses can selectively prevent the export of the MHC class I molecules to the cell surface, or induce their degradation once there. This might allow the infected cell to evade recognition by cytotoxic T-cells but would make it susceptible to being killed by NK cells. That is, viruses inhibit all protein synthesis in their host cells, so that synthesis of MHC class I proteins would be blocked in infected cells. The reduced level of MHC class I expression in infected cells would make them more susceptible to being killed by NK cells.

3.3 ADAPTIVE IMMUNITY

Unless indicated otherwise, we use the term "lymphocyte" to refer only to the antigen-specific lymphocyte. Lymphocytes allow a response against a vast array of antigens from various pathogens encountered in a person's life, and confer the important feature of immunological memory. These are possible through the highly variable antigen receptors on the surface of the lymphocyte. With the receptor, lymphocytes recognize and bind to antigens. Lymphocytes express a huge repertoire of receptors with antigen-binding sites against highly diverse antigens.

Among the billion or so lymphocytes circulating in the body, at any one time, there will be some lymphocytes capable of recognizing a given foreign antigen. A unique feature of the adaptive immune system is the capability of generating immunological memory, so that having been exposed once to an infectious agent, people make an immediate and stronger response against any subsequent exposure to it

[3]:371. That is, the adaptive immune system is capable of specifically recognizing a foreign antigen and generating immunological memory.

Lymphocytes are composed of B-cells and T-cells. Encountering antigens, lymphocytes become effector lymphocytes. The binding of an antigen to a BCR (B-cell receptor) activates the B-cell to secrete antibodies. A BCR is formed from a gene-encoding antibody secreted by the B-cell. Therefore, the BCR and the antibody have the same target.

Innate immune responses are the body's initial defense against infection, but these are nonspecific defenses. To effectively fight the wide range of pathogens, the lymphocytes of the adaptive immune system have evolved to recognize a great variety of different antigens from bacteria to viruses and other disease-causing organisms. An antigen is any molecule or part of a molecule that is specifically recognized by the highly specialized recognition proteins of lymphocytes. On B-cells, these proteins are the immunoglobulins (Igs) (see E3.7), which these cells produce in a vast range of antigen specifics, each B-cell producing immunoglobulins of a single specific. A membrane-bound form of immunoglobulin on the B-cell surface serves as the cell's receptor for an antigen, and is known as the B-cell receptor (BCR). A secreted form of immunoglobulin of the same antigen specific is the antibody produced by terminally differentiated B-cells: plasmablasts and plasma cells. The secretion of antibodies, which bind pathogens or their toxic product in the extracellular spaces of the body, is the main effector function of B-cells in adaptive immunity. The antigen-recognition molecules of T-cells are made solely as membrane-bound proteins, which are associated with an intracellular signaling complex and function only to signal T-cells for activation. These T-cell receptors (TCRs) are related to immunoglobulins both in their protein structure – both V-region (variable region) and C-region (constant region) – and in the genetic mechanism that produces their great variability. The T-cell receptor differs from the B-cell receptor in an important way: it does not recognize and bind antigens by itself, but instead recognizes short peptide fragments of protein antigens that are presented to them by proteins known as MHC molecules on the surface of host cells. Any given T-cell receptor is specific for a particular peptide bound to a particular MHC molecule.

3.4 FUNCTIONS OF ANTIBODIES

Antibodies are found in plasma, the fluid component of blood and extracellular fluid. Antibodies are Y-shaped molecules with two antigen-binding sites and a constant region or Fc region (fragment crystallizable region) (Figure 3.4). The constant region determines functional properties of the antibody, engaging with the effector mechanism and disposing of the antigen once recognized.

The ends of the two arms of the Y – the V-regions – are involved in antigen binding, and they vary in their detailed structure between different antibody molecules. The stem of the Y – the C-region – is far less variable and is the part that interacts with effector molecules and cells. These are five different classes of immunoglobulins, distinguished in their being constructed from C-regions that have different structures and properties. These are known as immunoglobulin M (IgM), immunoglobulin D (IgD), immunoglobulin G (IgG), immunoglobulin A (IgA), and immunoglobulin E

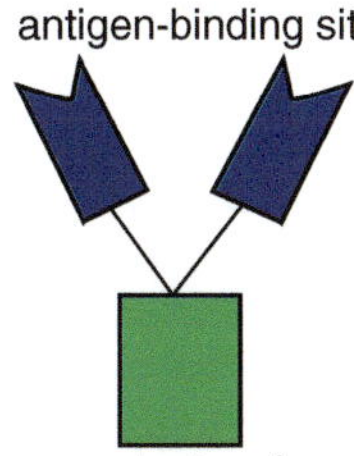

FIGURE 3.4 Antibody.

(IgE). The effector function of an antibody is defined by the structure of its heavy chain. All antibodies are constructed in the same way from paired heavy and light polypeptide chains so that two light chains are linked by disulfide bond to the tips of the arms of the Y whose trunk consists of two hinged heavy chains (Figure E3.5) [3]:140. That is, the antibody consists of the V-region involved in antigen binding and the C-region interacting with effector molecules and cells. All antibodies are constructed from paired heavy and light polypeptide chains. The function of an antibody is defined by the structure of heavy chain. There are five classes of immunoglobulins – IgM, IgD, IgG, IgA, and IgE – distinguished in their being constructed from C-regions.

The first way to protect against pathogens with antibodies is to bind antibodies to pathogens and block pathogens' access to cells which may be infected and destroyed. This is known as "neutralization," important for protection against viruses. This is the form of immunity elicited by most vaccines (Figure 3.5 left).

For bacteria, however, binding by antibodies is not sufficient to stop their replication. Hence, it is preferable that the antibody has a function to enable a phagocytic cell such as a macrophage or a neutrophil to ingest and destroy the bacterium. The second way is as follows. Most bacteria can evade because they have outer coats which are not recognized by pattern recognition receptors of phagocytes. But even antigens in the coats can be recognized by antibodies. Phagocytes have Fc receptors which admit the Fc region of antibody. The Fc receptors bind to Fc regions of antibodies. Phagocytes engulf bacteria bound to antibodies, leading to phagocytosis destroying bacteria (Figure 3.5 center panels). The treatment using antibodies in this way is known as "opsonization."

The third way is the complement activation. Complement is a factor of serum. Even without help of antibodies, complement is activated with microbial surfaces. When antibodies first bind to a bacterial surface, Fc regions of antibodies provide platforms of more effective complement activation than without antibodies. Complement components deposited on the bacterial surface can directly lyse bacterial membranes (Figure 3.5 right panels). The major function of complement is to enable phagocytes to engulf and destroy bacteria which are not recognized by the other ways. Most phagocytes express receptors called complement receptors binding to complement proteins deposited on bacterial surface and promoting phagocytosis.

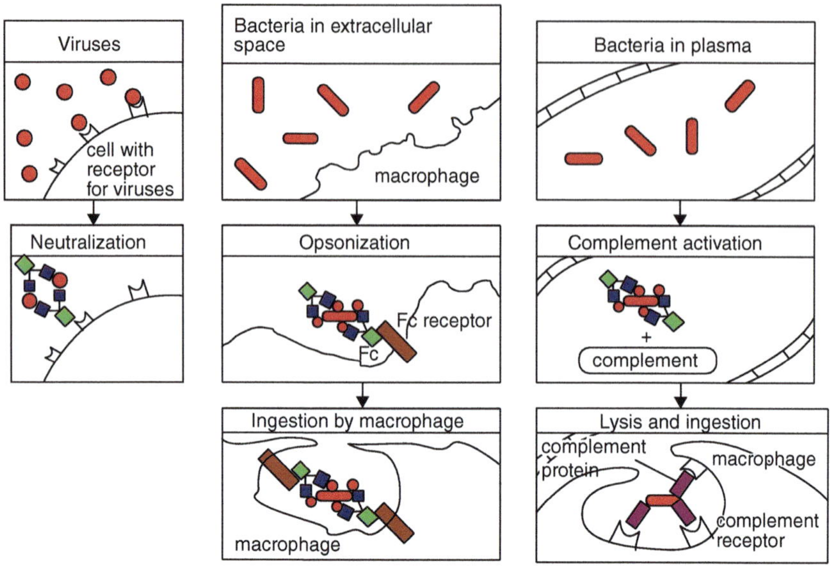

FIGURE 3.5 Three ways of host defense by antibodies.

The end result is that all pathogens are delivered to phagocytes for ingestion, degradation, and removal from body (Figure 3.5 bottom panels) [3]:28. Antibodies thus play the important role in protecting against infectious pathogens through neutralization and phagocytosis to ingest and degrade pathogens. That is, there are three ways to protect against pathogens with antibodies, such as neutralization, opsonization, and complement activation.

3.5 MHC (MAJOR HISTOCOMPATIBILITY COMPLEX)

Some bacteria and all viruses replicate inside cells. Their replication inside cells cannot be detected by antibodies because antibodies can only access blood and extracellular space. Destruction of intracellular invaders is the function of T-lymphocytes, the cells responsible for cell-mediated immune responses of adaptive immunity.

T-lymphocytes develop in the thymus. They are characterized by the types of T-cell receptors. The two main classes of T-cells express either a cell-surface protein called CD8 or another called CD4. These are important for a T-cell's function, because they help to determine the interactions between the T-cell and other cells. T-cells detect peptides derived from foreign antigens that are displayed by MHC molecules on a cell's surface. CD8 and CD4 function in antigen recognition by recognizing different regions of MHC molecules and by being involved in the signaling of the T-cell receptor that is engaged with its antigen.

Importantly, there are two main types of MHC molecules, called MHC class I and MHC class II. These have slightly different structures, but both have an elongated groove on the outer surface that can bind a peptide. The peptide becomes trapped in

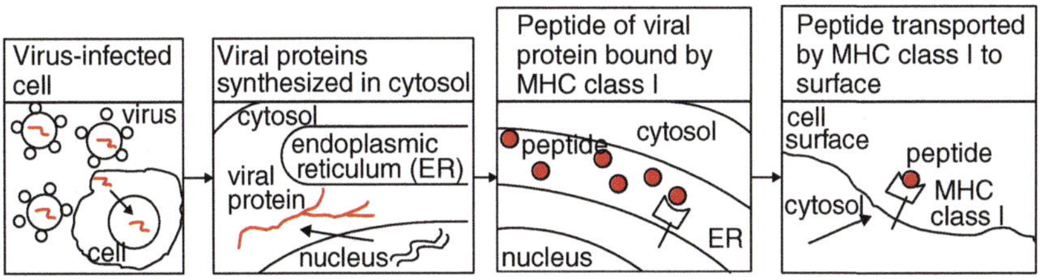

FIGURE 3.6 Peptide bound by MHC class I is expressed on the surface of a cell infected by viruses [3]:29.

this groove during the synthesis and assembly of the MHC molecule inside the cell, and the peptide-MHC complex is then transported to the cell surface and displayed to T-cells. CD8 recognizes a region of the MHC class I protein while CD4 recognizes a region of the MHC class II protein, distinguishing T-cells functionally. Therefore, CD8 T-cells selectively recognize peptides that are bound to MHC class I molecules, while CD4 T-cells recognize peptides bound to MHC class II molecules.

In cells infected with viruses, viral proteins are synthesized in the cytosol. Peptide fragments of viral proteins are transported into endoplasmic reticulum (ER), where they are bound by MHC class I molecules, which then deliver the peptides to the cell surface.

Most direct action of T-cells is cytotoxicity. Cytotoxic T-cells (Tc) are effector T-cells, acting against cells infected with viruses. Antigens derived from the virus multiplying inside the infected cell are displayed on the cell's surface (Figure 3.6), being recognized by antigen-receptors of cytotoxic T-cells. The cytotoxic T-cells can control infection by directly killing the infected cells before viral replication is complete and new viruses are released. Cytotoxic CD8 T-cells recognize epitope (antigenic determinant) buried within the antigen. Epitope peptides are presented by MHC (major histocompatibility complex) class I molecules, and serve as an important mechanism to defend against viral infection because MHC class I molecules are expressed in most cells of the body [3]:29. An MHC class I molecule without a peptide cannot be expressed on the surface of an antigen-presenting cell. In the case without MHC class I molecules on the surface of an antigen-presenting cell, T-cells cannot recognize the antigen-presenting cell and do not kill the cell, but NK cells kill the antigen-presenting cell as described in Figure 3.3 top. That is, viral proteins synthesized in the cytosol of cells infected with viruses are fragmented into peptide fragments, which are transported into the endoplasmic reticulum, where they are bound by MHC class I molecules which then deliver the peptides to the cell surface. Epitope peptides are recognized by antigen-receptors of cytotoxic T-cells, killing APC.

3.6 ANTIGEN-SPECIFIC RECEPTORS AND IMMUNE RESPONSE

3.6.1 Antigen-Specific Receptors of Adaptive Immunity

The innate immune system detects inflammatory stimuli by means of a relatively limited number of sensors, numbering fewer than 100 different types of proteins.

Antigen-specific receptors of adaptive immunity provide an infinite range of specificities, and are encoded in a finite number of genes. The basis for this extraordinary range of specificity was discovered in 1976 by Susumu Tonegawa, who was awarded the 1987 Nobel Prize [3]:15. Immunoglobulin variable regions are inherited as sets of gene segments, each encoding a part of the variable region of one of the immunoglobulin polypeptide chains. During B-cell development in the bone marrow, these gene segments are irreversibly joined by a process of DNA recombination to form a stretch of DNA encoding a complete variable region. A similar process of antigen-receptor gene arrangement takes place for the T-cell receptor genes during development of T-cells in the thymus.

Just a few hundred different gene segments can combine in different ways to generate thousands of different receptor chains. This combinatorial diversity allows a small amount of genetic material to encode a truly staggering diversity of receptors. During this recombination process, the random addition or subtraction of nucleotides at the junctions of the gene segments creates additional diversity known as junctional diversity. Diversity is amplified further by the fact that each antigen receptor has two different variable chains, each encoded by distinct sets of gene segments.

There are two critical features of lymphocyte development that distinguish adaptive immunity from innate immunity. Firstly, the process described earlier that assembles antigen receptors from incomplete gene segments is carried out in a manner that ensures that each developing lymphocyte expresses only one receptor specificity. Whereas the cells of the innate immune system express many different pattern recognition receptors and recognize features shared by many pathogens, the antigen receptor expression of lymphocytes is clonal, so that each mature lymphocyte differs from others in the specificity of its antigen receptor. Secondly, because the gene arrangement process irreversibly changes the lymphocyte's DNA, all its progeny inherit the same receptor specificity. Because this specificity is inherited by a cell's progeny, the proliferation of an individual lymphocyte forms a clone of cells with identical antigen receptors. That is, each lymphocyte expresses only one receptor specificity, whereas the cells of the innate immune system express many different pattern recognition receptors and recognize features shared by many pathogens, so that each mature lymphocyte differs from others in the specificity of its antigen receptor.

There are lymphocytes of at least 10^8 (one hundred million) different specificities in an individual human at any one time, comprising the lymphocyte receptor repertoire of the individual. These lymphocytes are continually undergoing a process similar to natural selection: only those lymphocytes that encounter an antigen to which their receptor binds will be activated to proliferate and differentiate into effector cells. This selective mechanism was proposed in the 1950s by Macfarlane Burnet who called this the clonal selection [3]:15.

3.6.2 Cytotoxic T-Cells

Cytotoxic T-cells recognize peptides bound to MHC class I molecules expressed on the infected cell's surface using a TCR (T-cell receptor), killing infected cells. When T-cells recognize antigens, they secrete cytokine IL-2 which enhances the

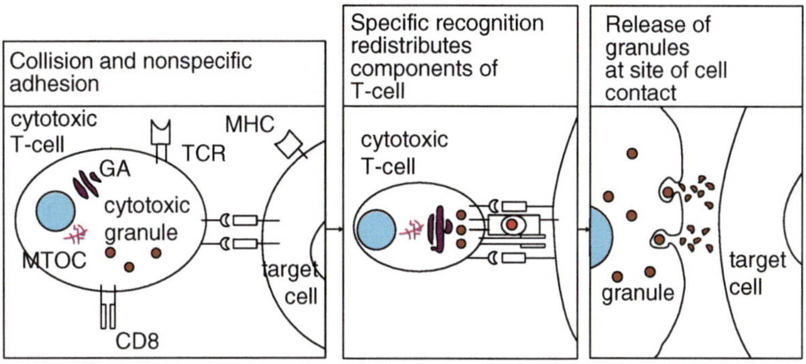

FIGURE 3.7 The cellular polarization of T-cells during specific antigen recognition [3]:383.

T-cell's clonal expansion and contributes to the T-cell's differentiation to effector status [3]:30.

We explain the T-cell's polarization using Figure 3.7, at the T-cell's contact to a target cell. When T-cells recognize a MHC class I-peptide complex expressed on the target cell's surface with a TCR, reorganization aligns the MTOC (microtubule-organizing center) present in cytosol of the T-cell at the site of contact. In turn, the MTOC aligns the secretory apparatus, including the GA (Golgi apparatus) secreting cytotoxic granules, toward the target cell. Granules are released to the antigen-presenting cells, killing cells infected with the antigen recognized by the T-cells. That is, when T-cells recognize MHC class I-peptide complex on the target APC, components of the T-cell such as MTOC and GA are redistributed, releasing cytotoxic granules toward the target APC.

We explain the process through which a T-cell kills a target cell, using Figure 3.8. When T-cells recognize a peptide-MHC complex on the target cell's surface by a TCR, T-cells release cytotoxic granules to the target cell. The cytotoxic granules consist of serglycin, granzyme, and perforin.

Through pores formed on the target cell's surface by perforin which is one of components of granule, granzymes enter into cytosol in the target cell. Then, granzymes cleave Bid, truncating it, and also cleave pro-caspase3, activating caspase3. Truncated Bid (tBid) binds to Bax on the surface of mitochondria, damaging mitochondrial outer membrane. From the damaged membrane, cytochrome c is released, enhancing apoptosis. Activated caspase3 cleaves inhibitory protein ICAD that binds to and inactivates CAD (caspase-activated deoxyribonuclease), freeing and activating CAD. Free CAD enters into the nucleus of the target cell, fragmenting DNA. Thus, the target cell with fragmented DNA is destroyed [3]:391.

3.6.3 Immune Response

In barrier tissues such as intestinal mucosae and respiratory mucosae, innate lymphoid cells (ILCs) lie in wait to eliminate a pathogen's spread through those poised barrier tissues. Most pathogens, however, develop the strategy to invade the innate

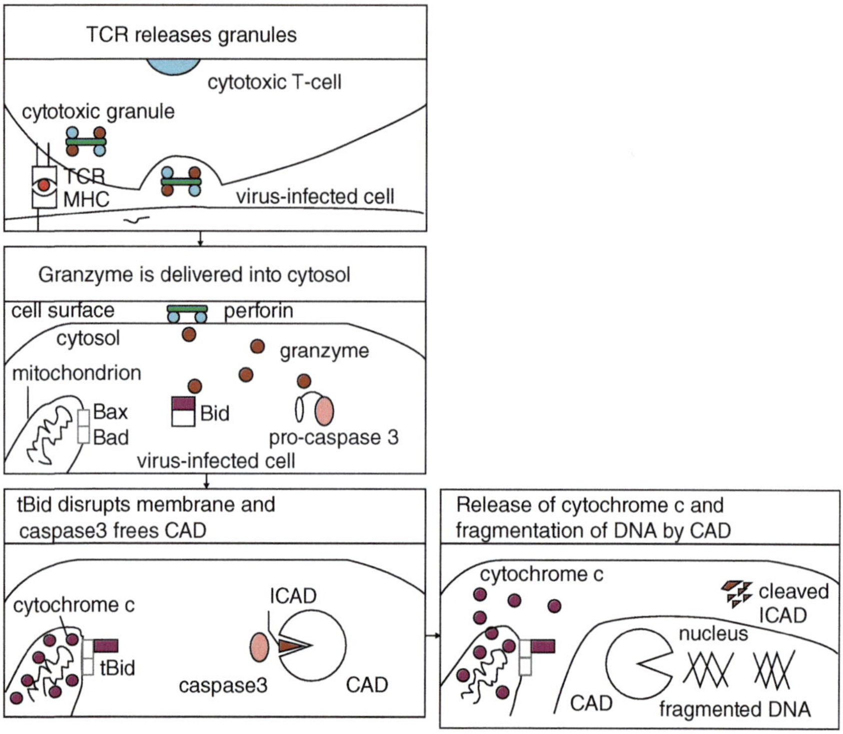

FIGURE 3.8 Granzymes released from cytotoxic granules induce apoptosis of the target cell [3]:391.

immune defense and establish infection. In these circumstances, the innate immune response introduces the adaptive immune response.

When a pathogen is encountered for the first time, the ILCs' primary immune response rapidly occurs in a few hours to days, activating naive lymphocytes to effector T-cells and initiating B-cells to secrete antibodies. But the adaptive response requires several days to weeks to fully mature, because of the rarity of antigen-specific precursor cells (Figure 3.9). The recognition specific to a pathogen encountered for the first time requires time for preparation. Once the preparation is finished, immunological memory is provided for a lifetime. Effector T-cells migrate to infected sites, effectively cleaving to the pathogen with antibodies that are pathogen-specific. When the infection is cleared, immunological memory provides protection against reinfection [3]:445. That is, the primary innate response against pathogens by ILCs rapidly occurs and introduces the adaptive immune response, activating naive lymphocytes to effector T-cells and initiating B-cells to secrete antibodies. The adaptive response requires several days to weeks to fully mature.

The immune response is described step wise as follows:

1. The infectious agent colonizes and its numbers increase as it replicates. The innate response is initiated immediately following detection of the pathogen.

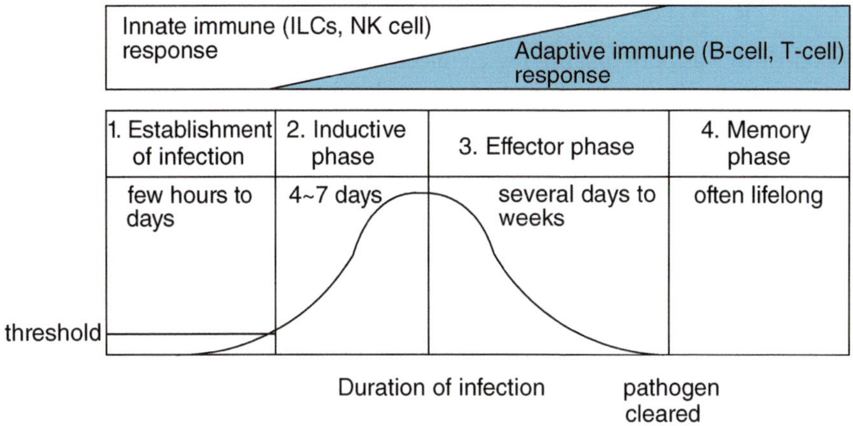

FIGURE 3.9 The adaptive immune response [3]:445.

2. When the number of pathogens exceeds the threshold dose of antigen required for an adaptive response, the adaptive response is initiated; the pathogen continues to grow, restrained by responses of the innate immune system. At this stage, immunological memory also starts to be induced.
3. After 4–7 days, effector cells and molecules of the adaptive response start to clear the infection.
4. When the infection has been cleared and the dose of antigen has fallen below the response threshold, the response ceases, but antibodies, residual effector cells, and immunological memory provide lasting protection against reinfection in most cases.

E3.1 USEFUL MICROORGANISMS

Microorganisms have two types of functions: corruption, which is harmful for humans, and fermentation, which is beneficial for humans. Fermentation is the process by which microorganisms decompose organic matter and change it to other matter. Microorganisms can decompose organic matter in circumstances without oxygen, producing energy.

In fermented foods, ingredients' taste and nutritive value are increased by functions of microorganisms. The pairings of fermented foods with their microorganisms are as follows:

- Yoghurt: lactic acid bacterium.
- Edible vinegar: acetobacter.
- Sake: malted rice.
- Wine: wine yeast.
- Bread: bread yeast.
- Beer: beer yeast.
- Soy sauce: soy sauce yeast.

E3.2 THE FATHER OF BACTERIOLOGY

Robert Koch was born of a mining engineer. He studied medicine at the University of Goettingen. He practiced medicine in a local town, researching bacteriology with light microscopy while a medical practitioner. In 1876, he discovered *Bacillus anthracis*, causing anthrax (means "charcoal" in Greek). In 1883, he discovered *Vibrio cholerae*, the bacterium behind cholera, and proved that infection to be caused by pathogens invading into body. In 1885, he joined as a professor of hygiene at Berlin University. In 1890, he developed tuberculin. In 1905, he was awarded the Nobel Prize in Physiology and Medicine [4].

E3.3 THE KITASATO INSTITUTE

In 1892, as the first director of a private infectious disease research institute, S. Kitasato made efforts toward eradication of infectious diseases, with his conviction linking the results of research to real treatment. The institute's management, however, was transferred to Tokyo University, leading to his anxiety in the execution of his intrinsic objective. Therefore, in 1914, he founded the Kitasato Institute, maintaining his conviction. He also founded the Faculty of Medicine of Keio Gijuku University prior to the Department of Medicine of the university [5].

E3.4 PROGENITORS OF BLOOD CELLS

The bone marrow contains multipotent hematopoietic stem cells, able to give rise to all classes of blood cells. In adult mammals, all of the cells shown in Figure E3.1 develop mainly in the bone marrow, except for T-lymphocytes, which develop in the

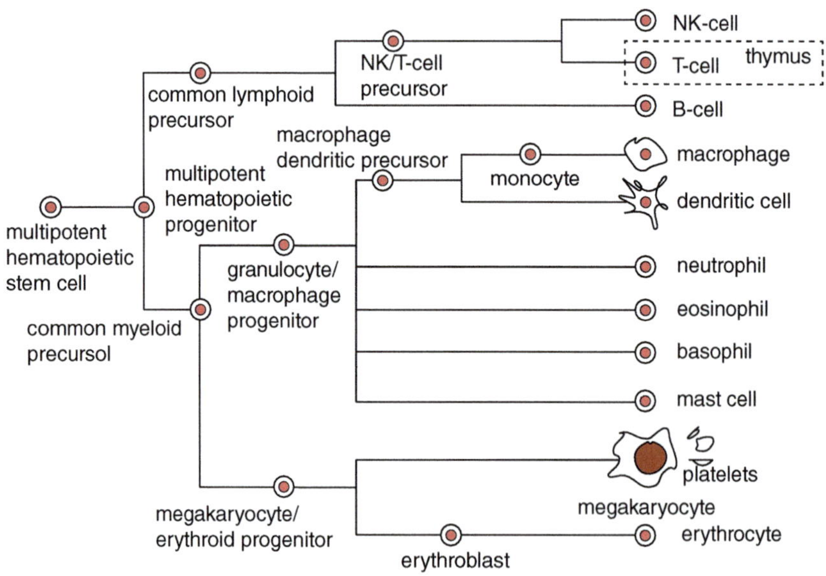

FIGURE E3.1 Development of all classes of blood cells [1]:1243.

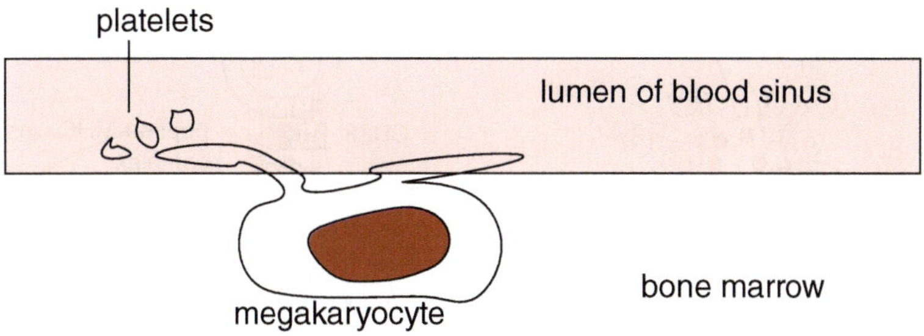

FIGURE E3.2 Platelets split off from a megakaryocyte's long processes [1]:1242.

thymus. The cells dedicated to phagocytosis, such as neutrophils, macrophages, and dendritic cells, develop from a common progenitor cell called a granulocyte/macrophage (GM) progenitor cell. Macrophages arise from differentiation of monocytes.

One megakaryocyte with enormous size produces about 10,000 platelets, which split off from long processes that extend through holes in the walls of an adjacent blood sinus (Figure E3.2) [1]:1242–1243.

E3.5 DENDRITIC CELLS

Dendritic cells, which are widely distributed in the tissues and organs of vertebrates, are crucially important components of the innate immune system. They express a large variety of PRRs (pattern recognition receptors), which enable dendritic cells to recognize and phagocytose (from Greek "phagein" meaning to eat) invading pathogens and their products, and to become activated in the process. The activated dendritic cells cleave the protein of the pathogen into peptide fragments, which bind to newly synthesized MHC proteins, which then carry the fragments to the dendritic cell surface. The activated dendritic cells then migrate to a nearby lymphoid organ such as a lymph node, where they present the peptide-MHC complexes to T-cells of the adaptive immune system, activating the T-cells to join in the battle against the specific pathogen. T-cells not activated by the antigen presented by dendritic cells exit from the lymph node via the cortical sinuses (Figure E4.1). T-cells activated by the antigen presented by dendritic cells start to proliferate and differentiate to effector T-cells, and exit from the lymph node (Figure E3.3).

The activated dendritic cells also secrete cytokines IL-12 and IL-18 influencing CD8 T-cells to produce IFN-γ, appropriate to fight the particular pathogen. In these ways, dendritic cells serve as crucial links between the innate immune system, which provides a rapid first line of defense against invading pathogens, and the adaptive immune system, which mounts a slower but more powerful and highly specific response to attack an invader [1]:1305. That is, dendritic cells thus serve as crucial links between the innate immune system and the adaptive immune system.

E3.6 CYTOKINES AND CHEMOKINES

Activating PRRs on sensor cells such as macrophages and neutrophils can directly induce effector functions in these cells, such as the phagocytosis and degradation of

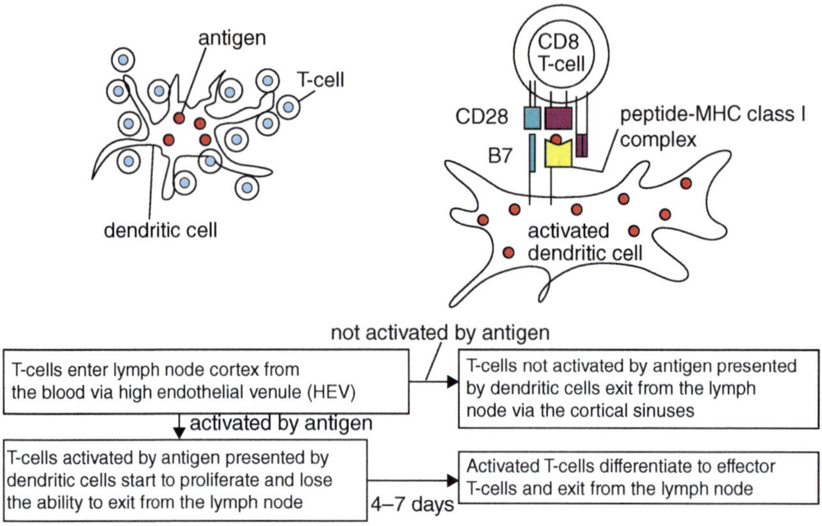

FIGURE E3.3 Dendritic cells activate T-cells [3]:351, 470.

bacteria they encounter. Sensor cells, however, serve to amplify the immune response by the production of inflammatory mediators. Two important categories of inflammatory mediators are the secreted proteins called cytokines and chemokines, which act in a manner similar to hormones to convey important signals to other immune cells. That is, sensor cells such as macrophages and neutrophils serve to amplify the immune response by production of inflammatory mediators. Two important inflammatory mediators are cytokines and chemokines, acting to convey important signals to other immune cells.

Cytokine is a term for any protein secreted by immune cells that affects the behavior of nearby cells bearing appropriate receptors. The response that a cytokine induces in a target cell is typically related to amplifying an effector mechanism of the target cell. Chemokines are a specialized subgroup of secreted proteins that act as chemoattractants, attracting cells bearing chemokine receptors, such as neutrophils and monocytes, out of the bloodstream and into infected tissue [3]:9–10. Figure E3.4 shows the signal protein TNF, for example, of cytokines. That is, a cytokine is a protein secreted by immune cells. Chemokines are a specialized subgroup of secreted proteins that act as chemoattractants, attracting cells bearing chemokine receptors, such as neutrophils and monocytes, at infected tissue.

The cytokines and chemokines released by activated macrophages act to recruit cells from the blood into infected tissues, a process known as inflammation that helps to destroy the pathogen. Inflammation increases the flow of lymph, which carries microbes or cells bearing their antigens from the infected tissue to nearby lymphoid tissues, where the adaptive immune response is initiated. Once adaptive immunity has been generated, inflammation also recruits these effector components to the site of infection. That is, the cytokines and chemokines released by activated macrophages enact inflammation which increases the flow of lymph that carries cells

Family	cytokine	Receptor	Producer cell	Action
TNF	TNF-α	TNFR-I	macrophage NK cell, T-cell	promotes inflammation
	LT-α	TNFR-I	T-cell, B-cell	killing
	CD40 ligand	CD40	T-cell	B-cell activation class switching
	Fas ligand	Fas (CD95)	T-cell	apoptosis

FIGURE E3.4 Signal protein TNF (tumor necrosis factor) [3]:813.

bearing their antigens from the infected tissue to nearby lymphoid tissues, where the adaptive immune response is initiated.

Inflammation is described clinically by the Latin words calor, dolor, rubor, and tumor, meaning heat, pain, redness, and swelling. Each of these features reflects an effect of cytokines or other inflammatory mediators on the local blood vessels. Heat, redness, and swelling result from the dilation and increased permeability of blood vessels during inflammation, leading to increased local blood flow and leakage of fluid and blood proteins into the tissues. Cytokines and complement fragments have important effects on the endothelium that lines blood vessels; the endothelial cells themselves also produce cytokines in response to infection. These alter the adhesive properties of the endothelial cells and cause circulating leukocytes to stick to the endothelial cells and migrate between them into the site of infection, to which they are attracted by chemokines. The migration of cells into the tissue and their local actions account for the pain. The main cell types seen in the initial phase of an inflammatory response are macrophages and neutrophils, the latter being recruited into the inflamed, infected tissue in large numbers. Macrophages and neutrophils are thus also known as inflammatory cells. The influx of neutrophils is followed a short time later by the increased entry of monocytes, which rapidly differentiate into macrophages, thus reinforcing and sustaining the innate immune response. Later, if the inflammation continues, eosinophils also migrate into inflamed tissues and contribute to the destruction of the invading microorganisms [3]:10. That is, the main cell types seen in the initial phase of an inflammatory response are macrophages and neutrophils, known as inflammatory cells.

E3.7 DIFFERENT CLASSES OF IMMUNOGLOBULINS

There are five different classes of immunoglobulins distinguished in their C-regions having different structures and properties. These are known as immunoglobulin M (IgM), immunoglobulin D (IgD), immunoglobulin G (IgG), immunoglobulin E (IgE), and immunoglobulin A (IgA), all of which can occur as transmembrane antigen receptors or secreted antibodies (Figure E3.5). In humans, IgG is found as four subclasses (IgG1, IgG2, IgG3, and IgG4), and IgA antibodies are found as two subclasses (IgA1 and IgA2). The different heavy chains defining these classes are known as isotypes and are designated by the lowercase Greek letters μ, δ, γ, ϵ, and α. The different heavy chains are encoded by different immunoglobulin C_H genes located in

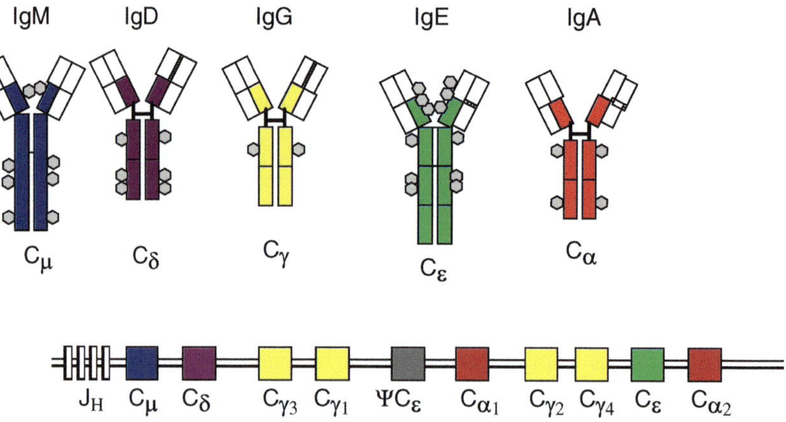

FIGURE E3.5 The immunoglobulin isotypes encoded by a cluster of immunoglobulin heavy-chain C-region genes [3]:192. The general structure of the main immunoglobulin isotypes is indicated, with each rectangle denoting an immunoglobulin domain. The constant region of the heavy chain for each isotype is indicated by the same color as the C-region segment that encodes it. IgM and IgE lack a hinge region but each contains an extra heavy-chain domain. Note the differences in the number and location of the disulfide bonds (black lines) linking the chains. The isotypes also differ in the distribution of N-linked carbohydrate groups, shown as hexagons.

a gene cluster that is 3' of the J_H segments. Figure E3.6 lists the major physical and functional properties of the different human antibody classes. IgM is the first class of immunoglobulin produced after activation of a B-cell. That is, the different heavy chains are encoded by different immunoglobulin heavy-chain C-region genes.

E3.8 ANTIGEN PRESENTATION FUNCTIONS

Pathogen-derived antigens are processed and presented for two distinct purposes. One is to induce the development of armed effector T-cells, and the other is to trigger the effector functions of these armed cells at sites of infection. MHC class I molecules bind peptides that are recognized by CD8 T-cells, and MHC class II molecules bind peptides that are recognized by CD4 T-cells. Nearly all somatic cells (except red blood cells) express MHC class I molecules. Consequently, the CD8 T-cell is primarily responsible for pathogen surveillance and cytolysis of somatic cells. CD8 T-cells are therefore an important mechanism in eliminating sources of new viral particles and bacteria that live only in the cytosol, and thus freeing the host from infection. By contrast, MHC class II molecules are expressed primarily only on cells of the immune system, and particularly by dendritic cells, macrophages, and B-cells. Thymic cortical epithelial cells and activated but not naive T-cells can express MHC class II molecules, which can also be induced on many cells in response to the cytokine IFN-γ. Thus, CD4 T-cells can recognize their cognate antigens during their development in the thymus, on a limited set of antigen-presenting cells, and on other somatic cells under specific inflammatory conditions. Effector CD4 T-cells comprise several subsets with different activities that help eliminate the pathogens. Importantly, naive

	Immunoglobulin								
	IgG1	IgG2	IgG3	IgG4	IgM	IgA1	IgA2	IgD	IgE
Heavy chain	γ_1	γ_2	γ_3	γ_4	μ	α_1	α_2	δ	ε
Half-life in serum (days)	21	20	7	21	10	6	6	3	2
Classical pathway of complement activation	++	+	+++	–	++++	–	–	–	–
Alternative pathway of complement activation	–	–	–	–	–	+	–	–	–
Binding to macrophage and phagocyte Fc receptors	+	–	+	– +	–	+	+	–	+
High-affinity binding to most cells and basophils	–	–	–	–	–	–	–	–	+++
Reactivity with staphylococcal protein A	+	+	– +	+	–	–	–	–	–

FIGURE E3.6 The physical and functional properties of human immunoglobulin isotypes. The relative activities of the various isotypes are compared for several functions, ranging from inactive (-) to most active (++++) [3]:193.

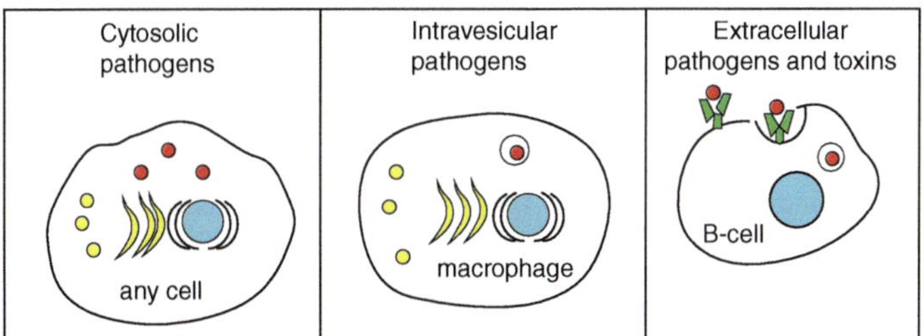

Degraded in	Cytosol	Endocytic vesicles (low pH)	Endocytic vesicle (low pH)
Peptides bind to	MHC class I	MHC class II	MHC class II
Presented to	Effector CD8 T-cells	Effector CD4 T-cells	Effector CD4 T-cells
Effect on presenting cell	Cell death	Activation to kill intravesicular bacteria and parasites	Activation of B-cells to secrete Ig to eliminate extracellular bacteria/toxins

FIGURE E3.7 Cells which acquire antigens from either the cytosolic or the vesicular compartments become targets of T-cell recognition [3]:215. Top, first panel: viruses and some bacteria replicate in the cytosolic compartment. Their antigens are presented by MHC class I molecules to activate killing by cytotoxic CD8 T-cells. Second panel: other bacteria and some parasites are taken up into endosomes usually by specialized phagocytic cells such as macrophages. Here, they are killed and degraded, or in some cases are able to survive and proliferate within the vesicles. Their antigens are presented by MHC class II molecules to activate cytokine production by CD4 T-cells. Third panel: proteins derived from extracellular pathogens may bind to cell-surface receptors and enter the vesicular system by endocytosis, illustrated here for antigens bound by the surface immunoglobulin of B-cells. These antigens are presented by MHC class II molecules to CD4 helper T-cells, which can then stimulate the B-cells to produce antibodies [3]:215.

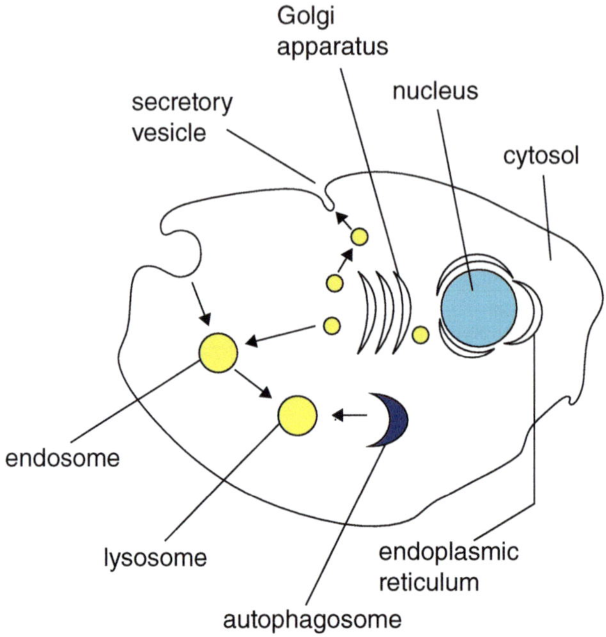

FIGURE E3.8 There are two categories of major intracellular compartments, separated by membranes [3]:215. One compartment is the cytosol, which communicates with the nucleus via pores in the nuclear membrane. The other is the vesicular system, which comprises the endoplasmic reticulum (ER), Golgi apparatus (GA), endosomes, lysosomes, and other intracellular vesicles. The vesicular system can be thought of as being continuous with extracellular fluid. Secretory vesicles bud off from the endoplasmic reticulum (ER) and are transported via fusion with the Golgi membrane to move vesicular contents out of the cell. Extracellular material is taken up by endocytosis or phagocytosis into endosomes or phagosomes, respectively. The fusion of incoming and outgoing vesicles is important both for pathogen destruction in cells such as neutrophils and for antigen presentation. Autophagosomes surround components in the cytosol and deliver them to lysosomes in a process known as autophagy [3]:215.

CD8 and CD4 T-cells can become armed effector cells only after encountering their cognate antigen once it has been processed and presented by activated dendritic cells. That is, antigen presentation has two purposes, one of which is to induce the development of armed effector T-cells, and the other of which is to trigger the effector function of these armed cells at sites of infection. MHC class I-peptide complex is recognized by CD8 T-cells, and MHC class II-peptide complex is recognized by CD4 T-cells. Nearly all somatic cells (except red blood cells) express MHC class I molecules. CD8 T-cells are important in cytolysis of somatic cells. By contrast, MHC class II molecules are expressed primarily only on cells of the immune system, particularly by dendritic cells, macrophages, and B-cells.

Peptides derived from the cytosol are transported into the endoplasmic reticulum (ER) and directly loaded onto newly synthesized MHC class I molecules on the same cell for recognition by T-cells. Because viruses and some bacteria replicate in the

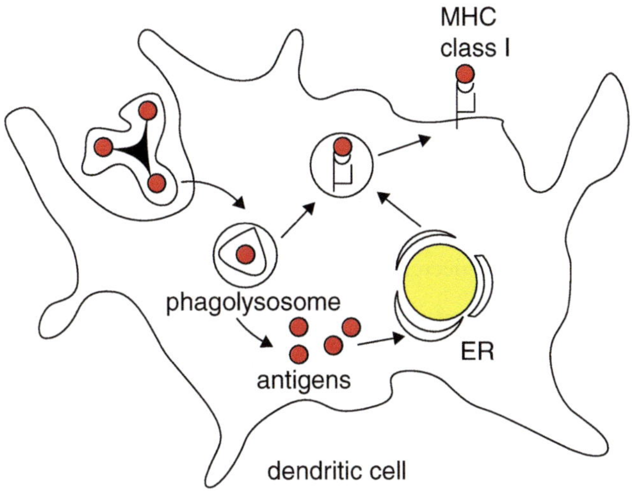

FIGURE E3.9 Cross-presentation of extracellular antigens on MHC class I molecules by dendritic cells [3]:216.

cytosol or in the contiguous nuclear compartment, peptides from their components can be loaded onto MHC class I molecules by this process (Figure E3.7, first upper panel). This pathway of recognition is referred to as direct presentation, and can identify both somatic and immune cells that are infected by a pathogen.

Certain pathogenic bacteria and protozoan parasites survive ingestion by macrophages and are able to replicate inside the intracellular vesicles of the endosomal-lysosomal system (Figure E3.7, second panel). Other pathogenic bacteria proliferate outside cells and can be internalized, along with their toxic products, by phagocytosis, receptor-mediated endocytosis, or macropinocytosis into endosomes and lysosomes (Figure E3.8), where they are broken by digestive enzymes. For example, receptor-mediated endocytosis by B-cells can efficiently internalize extracellular antigens through B-cell receptors (Figure E3.7, third panel). Virus particles and parasite antigens in extracellular fluid can also be taken up by these routes and degraded, and their peptides are presented to T-cells.

Some pathogens may infect somatic cells but not directly infect phagocytes such as dendritic cells. In this case, in order to process and present antigens to T-cells, dendritic cells must acquire antigens from exogenous sources. For example, to eliminate a virus that infects only epithelial cells, activation of CD8 T-cells will require that dendritic cells load MHC class I molecules with peptides derived from viral proteins taken up from virally infected cells. This exogenous pathway of loading MHC class I molecules is called cross-presentation, and is carried out very efficiently by some specialized types of dendritic cells (Figure E3.9). The activation of naive T-cells by this pathway is called cross-priming [3]:216. That is, if some pathogen were to infect somatic cells but not directly infect phagocytes such as dendritic cells, in order to present antigens to T-cells, dendritic cells must acquire antigens from exogenous sources. This exogenous pathway of loading MHC class I molecules is called cross-presentation carried out by dendritic cells.

4 Vaccination

4.1 HISTORICAL BACKGROUND

The two most important contributions to public health are sanitation and vaccination. These have markedly decreased death from infectious disease during the past 100 years.

Modern immunology grew from successes in vaccination by E. Jenner and L. Pasteur, i.e. vaccine against smallpox by E. Jenner and vaccine against cholera in chickens by L. Pasteur. The greatest success was the global eradication of smallpox announced by the WHO (World Health Organization) in 1979. The vision of vaccine researchers in the present generation is to level up the art of developing vaccine to modern drug design, that is to say, from experience to true pharmacology in the immune system.

The goal of vaccination is to generate long-lasting protective immunity, i.e. to generate protective immunity with immunological memory in order to eliminate pathogens by the innate and adaptive immune systems in the face of infection. The understanding that a single infection was sufficient to generate protective immunity was recorded in accounts of the Peloponnesian War more than 2,000 years ago. That is to say, during the war, two successive plague infections occurred in Athens. Greek historian Thucydides noted that people who survived the first infection did not contract the disease at the second infection. Knowledge of this type of relationship promoted the practice of variolation against smallpox, in which an inoculation of a small amount of material from a smallpox pustule was used to produce mild infection leading to protection against subsequent infection.

More than 1,000 years ago, smallpox was recorded in medical documents. In India and China, variolation seems to have been practiced many centuries before its introduction into the West. The variolation, however, was not necessarily mild, but was inappropriate according to the safety criteria of modern drugs. In E. Jenner's achievement, not only did infection with cowpox provide protective immunity against smallpox without risk of significant disease, but also experimental proof by intentional variolation was performed. E. Jenner named the process "vaccination" (from Latin "vacca" for cow).

Most modern vaccines offer protection by forming neutralization using antibodies. To use antibodies, however, is not fully protective against pathogens such as malaria, tuberculosis, and HIV (human immunodeficiency virus), for which additional activity such as cell-mediated immunity is requested. The present art of vaccine development does not effectively confer such immunity [3]:730.

4.2 CELL BIOLOGY OF INFECTION

In this section, we describe common tasks through which pathogens cause disease. The mechanisms through which pathogens cause disease are as diverse as

DOI: 10.1201/9781003780663-4

the pathogens themselves. Nonetheless, all pathogens must carry out certain common tasks: they must gain access to the host, reach an appropriate niche, avoid host defenses, replicate, and exit from the infected host to spread to an uninfected one [1]:1276.

The first step in infection is for the pathogen to gain access to the host. A thick covering of skin protects most parts of the human body from the environment. The protective boundaries of some other human tissue (eyes, nasal passages, respiratory tract, mouth, digestive tract, urinary tract, and female genital tract) are less robust. In the lungs and small intestine, for example, the barrier is just a single monolayer of epithelial cells. Nonetheless, all these epithelia serve as barriers to infection.

Wounds in barrier epithelia allow pathogens direct access to unoccupied niches within otherwise sterile host tissues. This avenue of entry requires little in the way of pathogen specialization, and many members of the normal flora can cause serious illness if they enter through such wounds. Staphylococci from the skin and nose or streptococci from the throat and mouth are two examples of opportunistic bacterial pathogens that are responsible for many serious infections resulting from breaches in epithelial barriers. The recent emergence of bacterial strains of staphylococcus that are resistant to the antibiotics commonly used for treatment (for example, methicillin-resistant *Staphylococcus aureus*, or MRSA, which infects up to 50,000,000 people worldwide) is of particular concern. Papillomaviruses, which cause warts and cervical cancer, also take advantage of breaches in epithelial barriers.

Primary pathogens, however, need not wait for a wound to gain access to their host. One efficient way for such a pathogen to cross the skin is to catch a ride in the saliva of a biting arthropod. A diverse group of bacteria, viruses, and protozoa have developed the ability to survive in insects and then use them as vectors to spread from one mammalian host to another. The plasmodium protozoan that causes malaria develops through several forms in its life cycle, including some that are specialized for survival in a human and others that are specialized for survival in a mosquito. Viruses that are spread by insect bites cause yellow fever and Dengue fever, as well as many kinds of viral encephalitis (inflammation of the brain). These viruses replicate in both insect cells and mammalian cells, as required for their transmission by an insect vector.

The efficient spread of a pathogen via an insect vector requires that an individual insect consumes a blood meal from an infected host and transfers the pathogen to a naive host. In a few striking cases, the pathogen alters the behavior of the insect so that its transmission to a new host is more likely. An example is the bacterium *Yersinia pestis* causing bubonic plague. It multiplies in the flea's foregut to form aggregated masses physically blocking the digestive tract; each repeated, but futile, attempt at feeding yields some of the bacteria in the foregut flushed into the bite site, thus transmitting the plague to a new host.

4.3 USING TWO TYPES OF VIRUSES

An antiviral vaccine uses two types of viruses: inactivated or dead virus, and live-attenuated virus. Between these, the live-attenuated virus is more effective. The reason is as follows. An inactivated or dead virus cannot replicate in a cell because the virus is not living. Hence, on the surface of infected cell, MHC-peptide complex is not expressed. In this case, a T-cell recognizing MHC-peptide complex is not necessary because of absent MHC-peptide complex. Therefore, immunity in this vaccine has no T-cell function and has no immunological memory.

Conversely, live-attenuated viral vaccines are generally far more potent: they elicit a greater number of effectors including the activation of CD4 T-cells and cytotoxic CD8 T-cells. CD4 T-cells help in shaping the antibody response, which is important for a vaccine's subsequent protective effect. Cytotoxic CD8 T-cells provide protection while infection by the virus itself is under way, and if maintained, may contribute to protective memory. Attenuated viral vaccines include the routine child vaccines in use for polio, measles, mumps, rubella, and varicella. Other attenuated live viral vaccines that are licensed for special circumstances or for use in high-risk populations include influenza, poxvirus (vaccinia), and yellow fever virus [3]:733. Activated CD8 T-cells confer protection against future infection and have immunological memory. As shown Figure 4.1, to produce an attenuated virus, the virus must first be isolated by growing it in cultured human cells. The adaptation to growth in cultured human cells can cause some attenuation in itself. In general, the virus is then adapted to growth in cells of a different species, for example, monkey cells, until it no longer grows well in human cells. The cultured virus is used for vaccination without risk of infecting human cells, helping to activate lymphocytes.

Activated CD4 T-cells differentiate into some subsets, one of which is T_{FH}. Protein antigens alone are unable to induce an antibody response in animals or humans who lack T-cells, and they are therefore known as thymus-dependent or TD antigens, and typically involve antigen-specific T-cell help. The T-cells involved are T_{FH} cells that reside in the lymphoid tissues and are not fully differentiated T_{H1}, T_{H2}, or T_{H17} effector cells. To receive T-cell help, the B-cell must display the antigen on its surface in a form that a T-cell can recognize. This occurs when the antigen bound by surface immunoglobulin on the B-cell is internalized and degraded within the B-cell and peptides derived from it are returned to the cell surface in a complex with MHC class II molecules. When the T_{FH} recognizes these peptide-MHC complexes, it provides the B-cell with signals that favor survival and induce proliferation. These signals include the activation of CD40 on B-cells by T_{FH} expression of its ligand, CD40L (CD154), and production of various cytokines by T_{FH} cells, including IL-21 (Figure 4.2). CD40

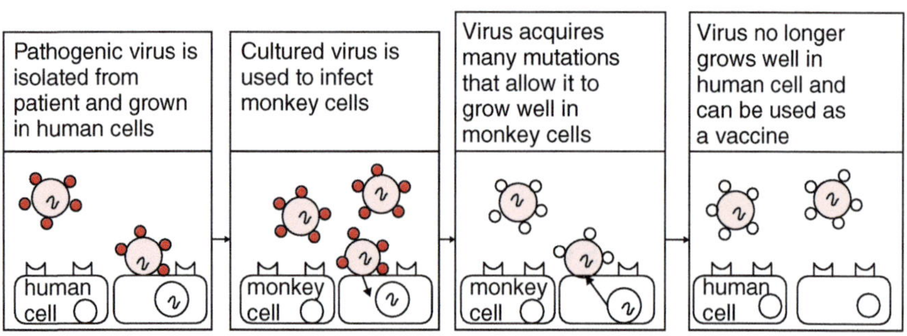

FIGURE 4.1 Viruses are traditionally attenuated in nonhuman cells [3]:733.

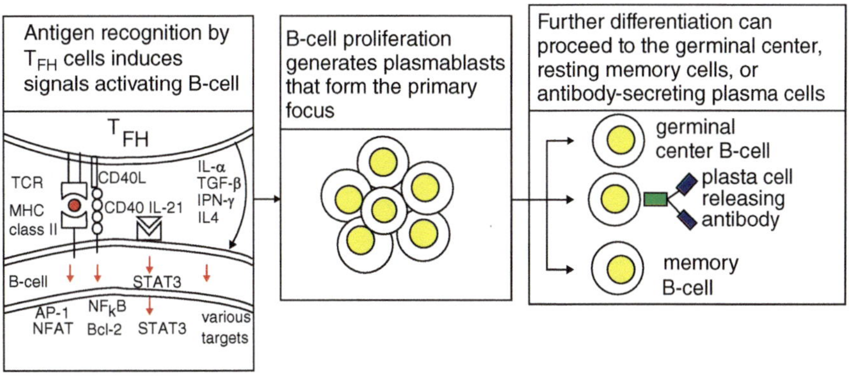

FIGURE 4.2 T_{FH} helps B-cells to secrete antibodies [3]:401.

signaling activates the non-canonical NF_kB pathway leading to the production of the cytokine IL-2 ([22]:285), and enhances B-cell survival by inducing the expression of anti-apoptotic molecules such as Bcl-2. IL-21 signaling activates STAT3 and enhances cellular proliferation and differentiation into plasma cells and memory B-cells. That is, T_{FH} is one of the subsets of activated CD4 T-cells. T_{FH} recognizes the peptide-MHC class II complex and provides the B-cell with signals which activate CD40 on B-cells by its ligand on T_{FH} and produce various cytokines by T_{FH} including IL-21. CD40 signaling activates the NF_kB pathway, producing cytokine IL-2. IL-21 signaling activates STAT3 and enhances proliferation and differentiation into plasma cells and memory B-cells.

Other cytokines produced by T_{FH} cells include IL-6, TGF-β, IFNγ, and IL-4, which provide signals that can regulate the type of antibody produced. After receiving these signals, activated B-cells begin to proliferate (Figure 4.2, second panel). Activated B-cells generate "plasmablasts" that form the "primary focus" (E4.4). Further differentiation can proceed to the "germinal center" (Figure E4.2), resting memory B-cell or "antibody-secreting plasma cells." Plasma cells release "antibodies." Thus, CD4 T-cells help B-cells to release antibodies (Figures 4.2) [3]:401. Thus, T_{FH} helps B-cells to secrete antibodies. That is, other cytokines such as IL-6, TGF-β, IFN-γ, and IL-4, produced by T_{FH}, provide signals which activate B-cells to proliferate and differentiate, leading to primary focus and germinal center.

4.4 MRNA VACCINE

The virus SARS-CoV-2 has spikes on its surface (Figure 4.4). When the spikes bind to receptors on surfaces of human cells, viruses invade human cells, initiating infection. The mRNA vaccine activates antigen-specific lymphocytes targeting spike proteins in order to have protection against future infection. As shown in Figure 4.4, in the process of making an mRNA vaccine, antigens are synthesized from mRNA coding spike protein, and coated with lipid for stability, being contained in mRNA vaccine. The mRNA vaccine is injected into a human body and a clinical trial is

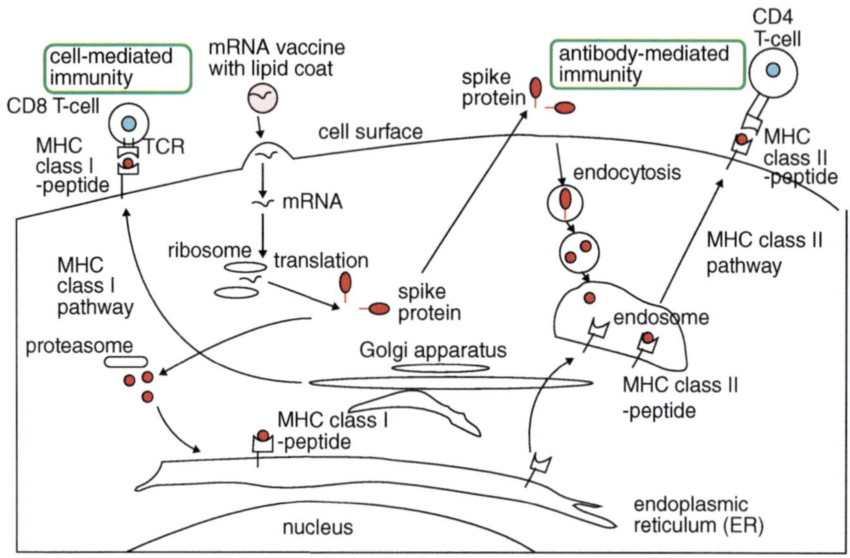

FIGURE 4.3 Immunity mechanism introduced by mRNA vaccine [6].

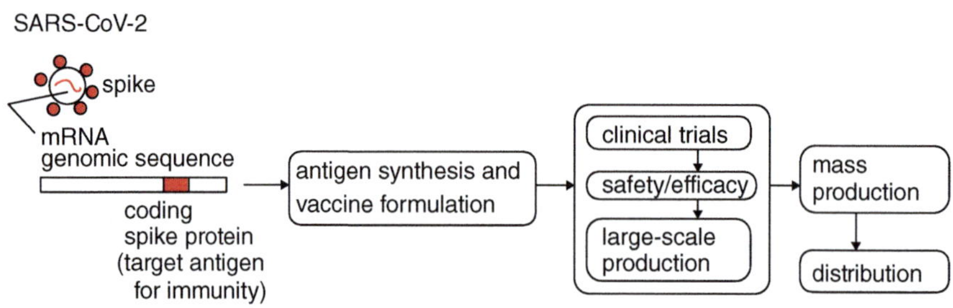

FIGURE 4.4 Vaccination for SARS-Cov-2 [7].

performed. If safety and efficacy in result satisfy the criteria, the vaccine is authorized, the is mass produced and delivered.

The function of an mRNA vaccine in a human cell upon being injected is described in the following using Figure 4.3. In the cytosol of the cell, spike proteins are produced from information from the mRNA by translation at the ribosome. The spike protein is treated as an antigen, whose fragment is produced as a peptide. A part of the peptide is bound in MHC class I-peptide complex and the other part of peptide is bound in MHC class II-peptide complex in ER (Endoplasmic reticulum) (Figure 3.6). These complexes are expressed on the surface of cell. MHC class I-peptide complex is recognized by CD8 T-cell, leading to cell-mediated immunity's function: protection against future infection and immunological memory.
On the other hand, MHC class II-peptide complex is recognized by CD4 T-cells, leading to antibody-mediated immunity's function: neutralization by antibody.

That is to say, CD4 T-cells activate B-cells, releasing antibodies which neutralize the antigen if encountered in the future to prevent toxicity from the antigen (Figure 4.2).

The novel coronavirus, called SARS-CoV-2 (severe acute respiratory syndrome coronavirus 2), was first reported in Wuhan, China in December 2019. SARS-CoV-2 has spread globally. The WHO proclaimed coronavirus disease 2019 (COVID-19) to be a pandemic in March 2020. As of November 2020, SARS-CoV-2 caused more than 54 million cases and 1.3 million deaths. In the US alone, it caused more than 11 million cases and more than 250,000 deaths. As of December 2020, BioNTech/Pfizer (E4.3) and Moderna reported that an mRNA vaccine has more than 90% protection efficacy based on analysis of phase III trials. BioNTech/Pfizer and Moderna were granted emergency authorization for a COVID-19 vaccine in December 2020 [7].

An mRNA vaccine has the following advantages:

1. An mRNA vaccine is safer than DNA vaccine: mRNA does not integrate into the host's genome, minimizing genotoxicity.
2. Compared with inactivated or live vector-based vaccines, production of an mRNA vaccine decreases biological contamination.
3. An mRNA vaccine induces cellular immune response with high efficacy.
4. The period for designing and producing an mRNA vaccine is short, meeting the need for a pandemic outbreak.

E4.1 E. JENNER

Edward Jenner was born on May 6, 1749 in Berkeley, Gloucestershire, England as the eighth of nine children. At the age of 14, he was apprenticed for seven years to Daniel Ludlow, a surgeon of Chipping Sodbury, South Gloucestershire, where he gained most of the experience needed to become a surgeon himself. In 1770, aged 21, E. Jenner became apprenticed in surgery and anatomy under surgeon John Hunter and others at St. George's Hospital, London. On May 14, 1796, Jenner tested his hypothesis by inoculating James Phipps, an eight-year-old boy who was the son of Jenner's gardener. He scraped pus from cowpox blisters on the hands of Sarah Nelmes, a milkmaid who had caught cowpox from a cow called Blossom, whose hide now hangs on the wall of the St. George's Medical School library. Phipps had no full-blown infection. The success of his discovery soon spread around Europe. Napoleon, who at the time was at war with Britain, had all his French troops vaccinated.

E4.2 INSTITUTE PASTEUR

Due to Louis Pasteur's successful development of the rabies vaccine, donations were made all around the world, resulting in the foundation of the Institute Pasteur in 1887 in Paris, with the objectives of developing science, medicine, and public health. Ever since its foundation, the Institute Pasteur has been following in the footsteps of Pasteur and conducting various activities as a public interest corporation of the private sector to contribute to the global health of mankind [8].

E4.3 K. KARIKÓ

In 1955, Katalin Karikó was born at Szolnok, Hungary. She received a doctoral degree from the University of Szeged. In 1990, while joining as a research assistant professor at the University of Pennsylvania, K. Karikó submitted her first grant application in which she proposed to establish mRNA-based gene therapy. In 2013, she was inaugurated as vice president at BioNTech RNA Pharmaceuticals and subsequently became a senior vice president in 2019. The technology of K. Karikó and Drew Weissman was used within a vaccine for COVID-19 that was produced jointly by Pfizer and BioNTech in 2020. K. Karikó and D. Weissman were awarded the Nobel Prize in Physiology and Medicine in 2023.

E4.4 PRIMARY FOCUS

Naive circulating B-cells enter lymph nodes from the blood by squeezing through the walls of specialized blood vessels called HEV (high endothelial venules) [3]:19 and are attracted by chemokines into the primary lymphoid follicle (Figures E4.1, E4.2). If these B-cells do not encounter an antigen in the follicle, they leave the lymph node via the efferent lymphatic vessel. On the other hand, B-cells that have bound to an antigen move to the border with the T-cell area, where they may encounter activated helper T-cells specific to the same antigen; these T-cells interact with the B-cells and activate them to start proliferation and differentiation into plasmablasts. A plasmablast is a B-cell in a lymph node that already shows some features of a plasma cell. Some B-cells activated at the T-cell–B-cell border migrate to form an aggregate of differentiating B-cells called the "primary focus" of antibody-secreting plasmablasts in the medullary cords (in lymph nodes), whereas others move back into the follicle, where they continue to proliferate and form a "germinal center." Germinal centers are sites of sustained B-cell proliferation and differentiation. Follicles in which germinal centers have formed are known as "secondary follicles." Within the germinal center, B-cells begin their differentiation into either antibody-secreting plasma cells

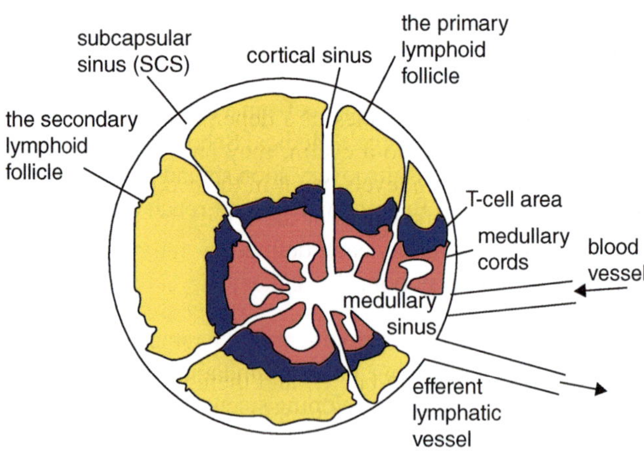

FIGURE E4.1 Lymph node [3]:351, 408.

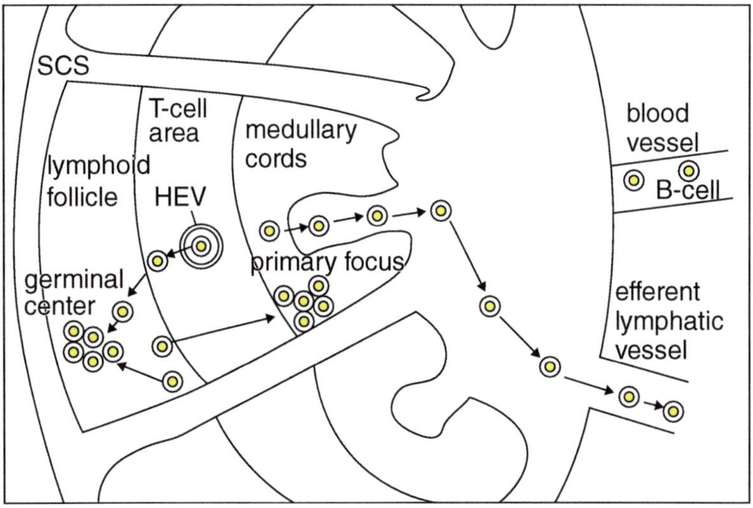

FIGURE E4.2 B-cells enter lymph nodes from the bloodstream via HEV and are attracted by chemokines into the lymphoid follicle [3]:408.

or memory B-cells. Plasma cells leave the germinal center and migrate to the medullary cords, or leave the lymph node via the efferent lymphatic vessel and migrate to the bone marrow [3]:408. The germinal center grows in size as the immune response proceeds, and then shrinks and finally disappears when the infection is cleared. Germinal centers are present for about 3–4 weeks after initial exposure. That is, naive B-cells entering lymph nodes from the blood through HEV are attracted into the primary lymphoid follicle. B-cells that have encountered the antigen move to the T-cell area where they are activated by T-cells to start proliferation and differentiation into plasmablasts. Some B-cells form a primary focus in the medullary cords, and others move back into the follicle, where they form a germinal center. Plasma cells leave the germinal center, migrate to medullary cords, or leave the lymph node via the efferent lymphatic vessel and migrate to the bone marrow. Germinal centers are present for about 3–4 weeks after initial exposure.

The primary focus reaction and the germinal center reaction differ in the quality of antibody that they produce. Plasmablasts in primary foci primarily secrete antibodies of the IgM isotype that offer some immediate protection. In contrast, B-cells in the germinal center reaction undergo several processes that produce antibodies that are more effective in eliminating infections. These processes in the germinal center include somatic hypermutation which alters the V-regions of immunoglobulin genes, and which enables a process called "affinity mutation," that selects the survival of mutated B-cells having a high affinity for the antigen [3]:408.

The antigen-binding region of a BCR (B-cell receptor) is known as the "variable region" or "V-region." The antigen recognition proteins of B-cells are the immunoglobulins (Igs) [3]:139. The antibody secreted by the B-cell is formed from the gene encoding BCR. Therefore, the target of the antibody is the same as the antigen recognized by the BCR.

5 Treatment for Cancer

5.1 WHY DO CANCER CELLS OCCUR?

For cancers known to be due to a specific external cause, the disease does not become apparent until long after exposure to the causal agent. For example, the incidence of lung cancer does not begin to rise steeply until after decades of heavy smoking. Similarly, industrial workers exposed to chemical carcinogens do not develop the cancer characteristic of their occupation until ten, twenty, or even more years after the exposure. Therefore, the concept that the development of cancer requires the gradual accumulation of mutations in a number of different genes helps to explain the tumor progression. The initial mild disorder of cell behavior gradually evolves and eventually becomes full-blown cancer [1]:1095. The fact that the incidence rate of cancer begins to rise steeply after 40 years old helps to explain cancer progression due to mutation accumulation (Figure 5.1).

Cancer depends on an accumulation of inherited changes in cells. Genes whose changes contribute to causation or evolution of cancer are called "cancer-critical genes." Cancer-critical genes are grouped into two classes. In one of the classes, mutations that bring about a gain of function can drive a cell toward cancer. Such genes are called "protooncogenes," whose overactive or overexpressive forms are called "oncogenes."

In the other class, mutations that bring about a loss of function can contribute to cancer cell growth. Such genes are called tumor-suppressor genes. In Figure 2.12, active Rb makes E2F protein inactive, and holds the cell cycle. However, when Rb loses such function by mutation, even in cases of no active G_1-CDK, E2F protein becomes active because it is not blocked by Rb and the cell cycle becomes overactive, leading to cancers. Hence, Rb is a tumor-suppressor gene. As shown in Figure 2.13, when DNA is damaged by an X-ray, p53 has a function to stop DNA synthesis. But the loss of p53's function due to mutation leads to proliferation of damaged DNA, resulting in cancer. Therefore, p53 is a tumor-suppressor gene. Cancer-critical genes are classified into two classes, the one of which is the oncogene such as Ras (E5.1), and the other of which is the tumor-suppressor gene such as Rb and p53 (E5.2, E5.3).

5.2 VIRUSES ASSOCIATED WITH CANCER

A small but significant proportion of human cancer is derived from the mechanism that involves viruses. Papillomaviruses are a prime example of DNA tumor viruses. This is important as a cause of carcinoma of the uterine cervix. A vaccine has now been developed for protection against infection with papillomaviruses. It is recommended that girls before puberty be vaccinated to reduce their risk of developing cervical cancer [1]:1131.

DOI: 10.1201/9781003780663-5

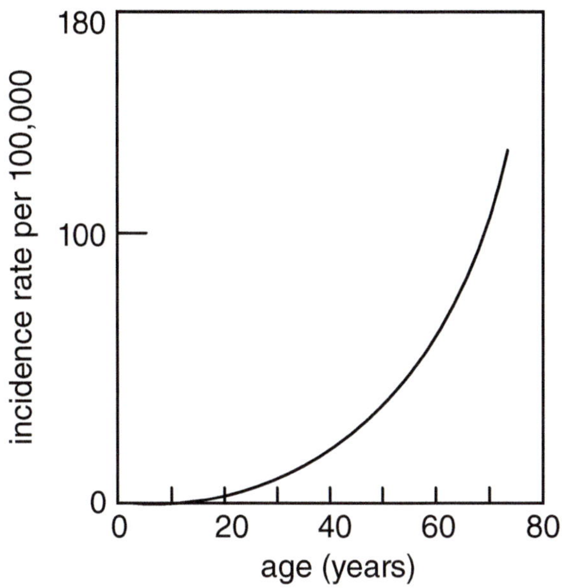

FIGURE 5.1 The incidence rate of cancer [1]:1095.

5.3 GENETIC INSTABILITY IN CANCER

Cancer therapies necessitate the adoption of a molecular peculiarity of cancer cells as an advantage. The one property is genetic instability resulting from a fault in chromosome maintenance for the DNA repair. The DNA repair necessitates Brca1, Brca2, and PARP (poly ADP-ribose polymerase). For the repair of DNA double-strand breaks, tumor-suppressor genes, i.e. Brca1 and Brca2, are necessary, whereas for the repair of single-strand breaks, another mechanism involving an enzyme called PARP is necessary. Realizing this DNA repair mechanism lead to a striking discovery. Drugs blocking PARP activity have been found to kill Brca-deficient cells with extraordinary selectivity. The drugs adopt the advantage of a cancer cell's genetic instability due to loss of Brca1 and Brca2 necessary for the DNA repair. These drugs are called PARP inhibitors [1]:1133–1135. Mutations in tumor-suppressor genes Brca1 and Brca2 lead to breast cancer [1]:281.

Knocking out a part of the pathway for DNA repair does not lead to a wretched cell. But knocking out the entire pathway for DNA repair derives a cell with severe damage, because alternative DNA repair cannot be available. When cancer that by mutation loses the Brca1 and Brca2 necessary for the repair of double-strand breaks, also loses PARP activity by drugs inhibiting the PARP necessary for the repair of single-strand break, all the possible repairs of breaks are knocked out and DNA breaks increase, leading to cancer cell death, because of fragmented DNA. A cell with DNA damage thus necessitates Brca1, Brca2, and active PARP for DNA repair. Hence, a cell with DNA damage, losing Brca1, Brca2, and active PARP, is led to cell death because of the impossibility of DNA repair. Thus, treatment with drugs that inhibit activity of PARP can destroy cancer that has already lost Brca1 and Brca2 due to mutation.

5.4 THE MOST COMMON FATAL DISEASE

In industrialized countries, one of the three leading causes of death is cancer. The others are infectious disease and cardiovascular disease, treatments for which continue to become improved, increasing the average life expectancy.

On the other hand, cancer is likely to become the most common fatal disease. Cancer is caused by uncontrollable growth of a mutated cell's progeny. The major problem is to control metastasis. Therefore, cancer therapy requires the removal and destruction of malignant cells. The method of performing this objective is to induce the immune response discriminating between cancer and normal cells. It is a similar method to vaccination.

Although immunological approaches have been used to treat cancer for more than one century, it is only in the past decade that real promise has been shown. The most important conceptual advance is the integration of conventional approaches, such as surgery and chemotherapy, with immunotherapy [3]:716. In 1908, for the first time, Nobel prize winner Paul Ehrlich proposed that the immune system could be used to treat cancer, suggesting that antibodies could be used to deliver toxins to cancer cells.

In 1950, immune surveillance was formulated as a hypothesis by Frank MacFarlane Burnet and Lewis Thomas. Because later it became clear that the relationship between the immune system and cancer was considerably complex, the hypothesis was modified, and began to consider three phases of cancer growth. The first is the elimination phase during which the immune system detects cancer cells and destroys them, referred to as immune surveillance as Figure 5.2. Incomplete elimination leads to the next equilibrium phase during which cancer cells receive change or mutation aiding their survival of pressure from the immune system.

In this phase, the process known as cancer immunoediting shapes properties of surviving cancer cells. In the final escape phase, cancer cells acquire the capability of eluding the immune system, growing unimpeded until becoming clinically detectable (E5.4) [3]:717.

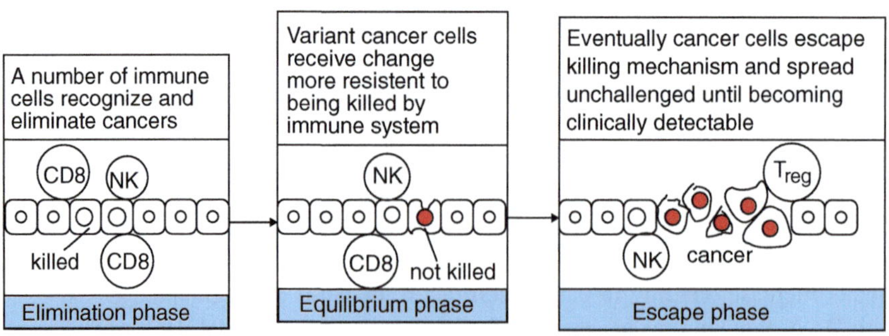

FIGURE 5.2 Malignant cells can be controlled by immune surveillance [3]:717.

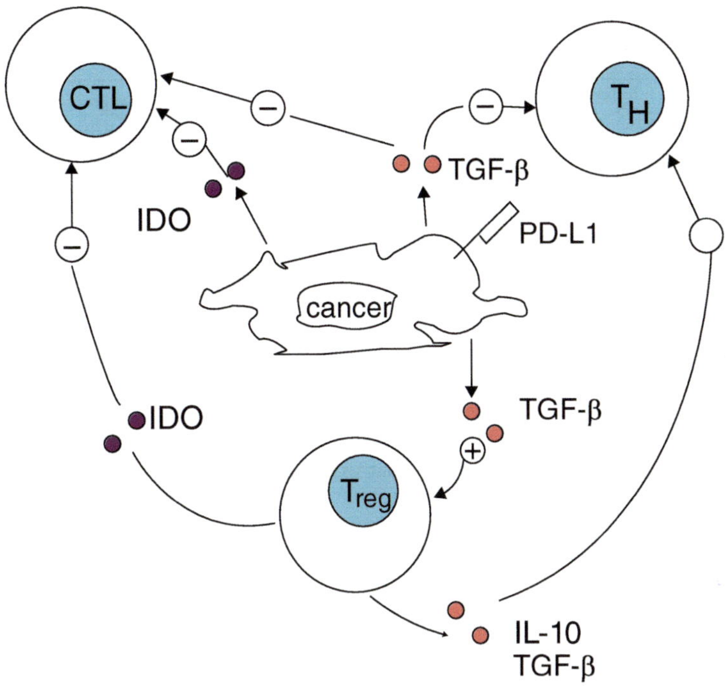

FIGURE 5.3 Cancer's escape mechanism avoiding immune recognition [3]:719.

5.5 CANCER'S ESCAPE FROM IMMUNITY

Because cancer cells tend to be genetically unstable, their clones lose the function of MHC class I expression. In this case, cancer cells are not recognized by T-cells, escaping the immune system. As shown in Figure 5.3, cancers make immunosuppressive cytokines propagating important information to other cells. TGF-β (transforming growth factor-β) secreted from cancer tends to suppress the T-cell response necessary to control cancer growth. TGF-β induces the progress of T_{reg} which is found in various cancers, specifically expanding in response to the cancer antigen. Several cancers produce immunosuppressive cytokine IL-10 that reduces dendritic cell activity, inhibiting T-cell activation.

Some cancers express PD-L1 (programmed death ligand-1) on the surfaces, which is the ligand to the inhibitory receptor PD-1 on the surface of an activated T-cell. When PD-L1 binds to PD-1, the T-cell is inactivated. Furthermore, cancer makes the enzyme IDO (indoleamine 2,3dioxygenase) (E5.6), which suppresses the local immune response (Figure 5.3).

5.6 NK (NATURAL KILLER) CELLS KILLING CANCER CELLS

When cancer cells express MHC class I-peptide complex on their surfaces, CTLs (cytotoxic T-lymphocytes) recognize cancer cells, killing cancers. However, the

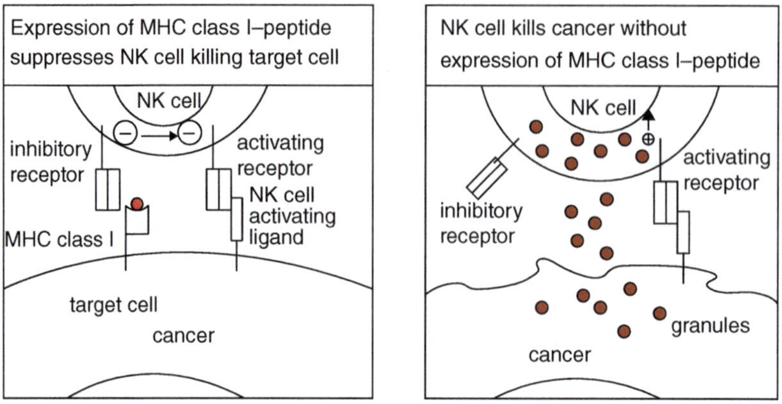

FIGURE 5.4 Expression of MHC class I-peptide complex suppresses NK cells killing the target cell. But NK cells kill cancer without expression of MHC class I-peptide complex [3]:127.

presence of MHC class I-peptide complex makes an inhibitory receptor on an NK cell's surface release an inhibitory signal to the NK cell, not killing the cancer (Figure 5.4 left). Because cancer is genetically unstable, its clone may not express MHC class I-peptide complex. In that case, although cancer is not killed by CTLs, there is no inhibitory signal to the NK cell, killing cancer. Thus, cancer escaping CTLs cannot escape NK cells, which release granules inducing apoptosis in the target cell (Figure 5.4 right) [3]:127.

5.7 CTLs (CYTOTOXIC T-LYMPHOCYTES) KILLING CANCER CELLS

Both CD8 T-cells and CD4 T-cells are important to achieve immunological control of cancer. CD8 T-cells directly kill cancer. On the other hand, CD4 T-cells play a role in activating CD8 T-cells by producing abundant IL-2 and helping to drive CD8 T-cell proliferation. This may in turn activate the CD8 T-cell to make its own IL-2 (Figure 5.5), establishing immunological memory. A subset T_{H1} of CD4 T-cells secretes cytokine TNF-α (tumor necrosis factor-α) (Figure E3.4) activating macrophages, killing cancer cells [3]:384, 722.

Cancer-specific antigens become targets of cancer-specific T-cell responses. For example, β-catenin is an antigen of melanoma, WT1 is an antigen of leukemia, and HER2/neu is an antigen of breast cancer [3]:721–722.

5.8 ANTIBODY CONTROLLING CANCER

To destroy cancer, an antibody needs to target a cancer-specific antigen expressed on the surface of cancer cells. The cancer-specific antigens are shown in Figure 5.7. Antibody-toxin conjugate or antibody-radionuclide can destroy cancer targeted by the antibody (Figure 5.6).

First occurrence of antibody controlling cancer is shown in Figure 5.6 left. When the Fc of the antibody binding the cancer-specific antigen is bound to

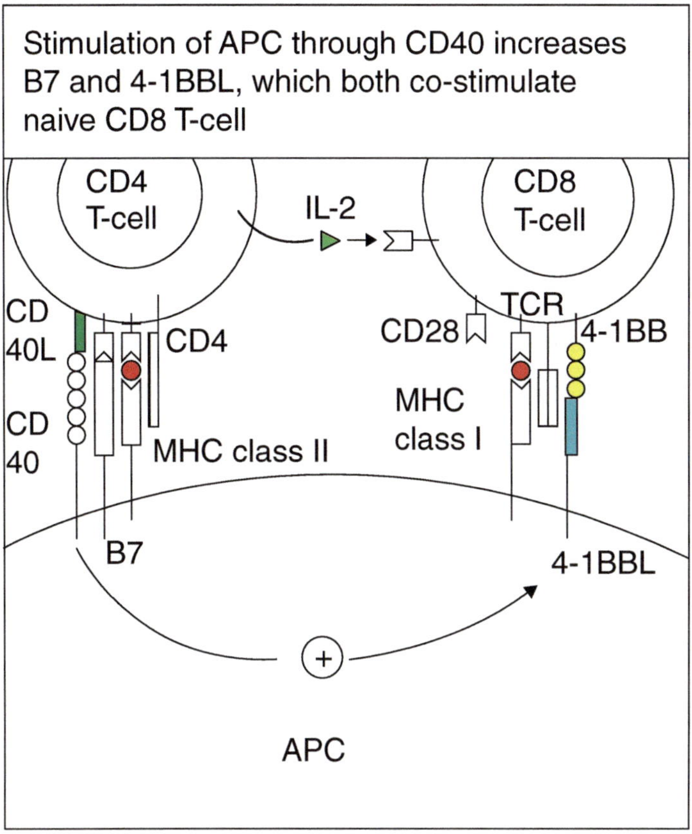

FIGURE 5.5 CD4 T-cells stimulated by APC activate CD8 T-cells [3]:372.

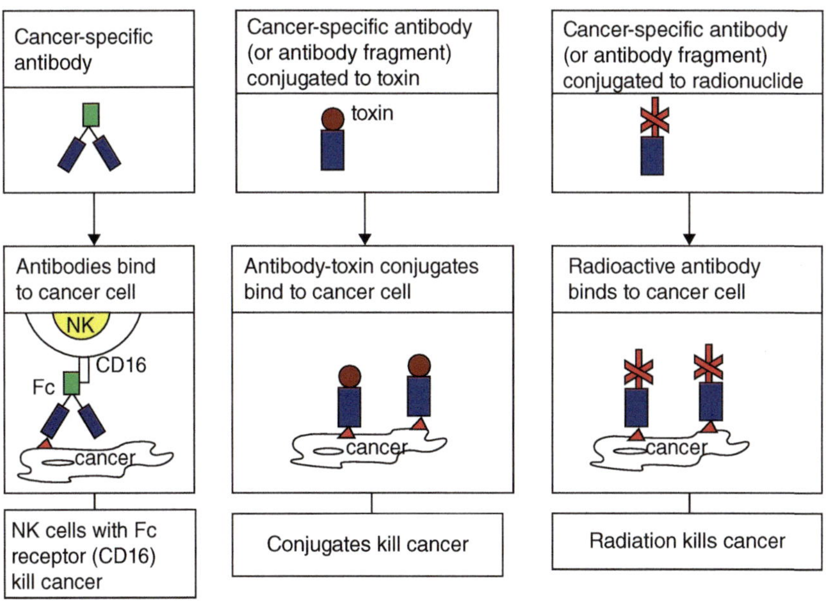

FIGURE 5.6 Antibodies that recognize cancer-specific antigens have been used to help eliminate cancers.

Tumor tissue origin	Type of antigen	Antigen	Tumor type
Lymphoma/ leukemia	Differentiation antigen	CD5 Idiotype CD52 (Campath-1H)	T-cell lymphoma B-cell lymphoma T-cell and B-cell lymphoma/leukemia
	B-cell signaling receptor	CD20	Non-Hodgkin's B-cell lymphoma
Solid tumors	Cell-surface antigens Glycoprotein	CEA, mucin-1	Epithelial tumors (breast, colon, lung)
	Carbohydrate	Lewis CA-125	Epithelial tumors Ovarian carcinoma
	Growth factor receptor	Epidermal growth factor receptor (EGFR) HER2/neu IL-2 receptor Vascular endothelial growth factor (VEGF)	Lung, breast, pancreatic, and head and neck tumors Breast, ovarian tumors T-and B-cell tumors Colon cancer Lung, prostate, breast
	Stromal extracellular antigen	FAP-α Tenascin Metalloproteinases	Epithelial tumors Glioblastoma multiforme Epithelial tumors

FIGURE 5.7 Examples of cancer-specific antigens that have been targeted by antibodies [3]:725.

an Fc receptor (CD16) of an NK cell, activated NK cells kill cancer cells. It was reported that the striking improvement in survival of breast cancer patients treated with antibody trastuzumab (Herceptin) targeting the antigen HER2/neu was accomplished [3]:724. HER2/neu is overexpressed in about one quarter of breast cancer patients.

Monoclonal antibodies can also be conjugated to chemotherapeutic drugs or to a radioisotope. In the case of a drug-linked antibody, the binding of the antibody to a cell surface antigen concentrates the drug to the site of the tumor. After internalization, the drug is released into the endosomes and exerts its cytostatic or cytotoxic effect. For example, the antibody trastuzumab has been linked to the cytotoxic agent mertansine, a drug that inhibits the assembly of microtubules, in a conjugate called T-DM1. Since HER2 is overexpressed only in cancer cells, T-DM1 selectively delivers the toxin specifically to cancer cells. Another drug-antibody conjugate, i.e. brentuximab vedotin, links an antiCD30 antibody with a different microtubule inhibitor and is approved for certain forms of relapsed lymphomas [3]:725–726. In the case of a radioisotope-linked antibody, monoclonal antibodies linked to radioisotopes have been successfully used to treat refractory B-cell lymphoma, using anti-CD20 antibodies linked to yttrium-90 (ibritumomab tiuxetan).

Monoclonal antibodies are defined as antibodies with the property that all the antibody molecules are identical in structure, including their antigen-binding site and isotype. The technique producing monoclonal antibodies was developed by G. Kohler and C. Milstein [3]:758.

5.9 CHEMOTHERAPY WITH DRUGS

One strategy for treating cancer is to attack cancer-expressing oncogenes, with s designed to specifically block the function of protein produced from the oncogene. Cancer that has undergone oncogene mutation once often further mutates, leading to hyperactivity of the initial oncogene. Blocking activity of the oncogenic protein may kill cancer cells without damaging neighboring normal cells.

Chronic myelogenous leukemia (CML) is usually associated with chromosomal translocation, visible as the Philadelphia chromosome (E5.7). This results from chromosome breakage and rejoining at the sites of two specific genes, Abl and Bcr (E5.7). The fusion of these two genes creates a hybrid gene called Bcr-Abl, coded for a chimeric protein consisting of the N-terminal fragment of Bcr fused to the C-terminal portion of Abl [1]:1135. Abl is a tyrosine kinase involved in cell signaling. The substitution of the Bcr fragment for the normal N-terminus of Abl makes it hyperactive, so that it stimulates inappropriate proliferation of the hemopoietic precursor cells that contain it and prevents these cells from dying by apoptosis as many of them would normally do. As a result, excessive numbers of white blood cells accumulate in the bloodstream, producing CML [1]:1135.

CML can be treated with a drug targeting and blocking the chimeric Bcr-Abl protein. This drug, called imatinib (trade name Gleevec), inhibits activity of tyrosine kinase, blocking Bcr-Abl as shown in Figure 5.8.

A dramatic response to the drug was shown. In over 80% of patients, cells carrying the Philadelphia chromosome disappeared and the response appeared relatively durable after years of continuous treatment [1]:1135.

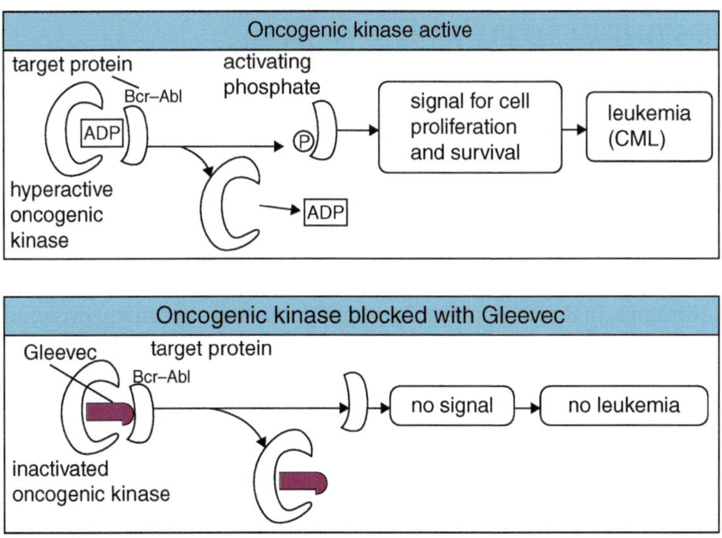

FIGURE 5.8 The anti-cancer drug imatinib (Gleevec) blocks the activity of oncogenic kinase Bcr-Abl and halts CML [1]:1136.

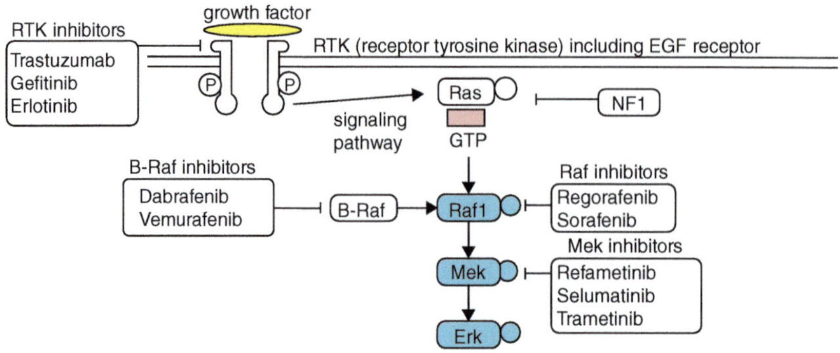

FIGURE 5.9 Some anti-cancer drugs and drug targets in the Ras-MAP kinase signaling pathway [1]:1137.

We can treat cancer using anti-cancer drugs to attack cancer cells. Chemotherapy uses anti-cancer drugs to inhibit protein on signaling pathways that promote the cell cycle of cancer cell, inhibiting activity of the protein and preventing cancer's proliferation. Drugs with this action are shown in Figure 5.9. About 25% of breast cancers, for example, express unusually high levels of the HER2 protein, a receptor tyrosine kinase related to the EGF receptor. A monoclonal antibody called trastuzumab (trade name Herceptin), which binds to Her2 and inhibits its function slows the growth of breast tumors in humans that overexpress Her2 and it is now an approved therapy for these cancers [1]:1137. Drugs that are antibodies end in "-mab," while those that are small molecules end in "-nib."

5.10 IPS (INDUCED PLURIPOTENT STEM) CELLS FOR DRUG DISCOVERY

In this section, we describe that iPS and transdifferentiated cells promise to be used for investigation of disease mechanisms and drug discovery.

A large part of the excitement surrounding ES and iPS cells and the technology of transdifferentiation comes from the prospect of using the artificially generated cells for tissue repair. It is beginning to seem that virtually any type of tissue might be replaceable, allowing treatment of degenerative diseases that have previously had no cure. Research in this area is moving rapidly, but there are many difficulties to be overcome.

With the advent of iPS cells and direct transdifferentiation, at least one major hurdle has been surmounted, in principle at least: the problem of immune rejection. Embryonic stem (ES) cells, because they are created from early embryos that generally come from unrelated donors, will never be genetically identical to the cells of the patient receiving the transplant. The transplanted cells and their progeny are therefore liable to rejection by the immune system. Both iPS and transdifferentiated cells, in contrast, can be generated from a small sample of the patient's own tissue and so should escape immune attack when transplanted back into the same individual.

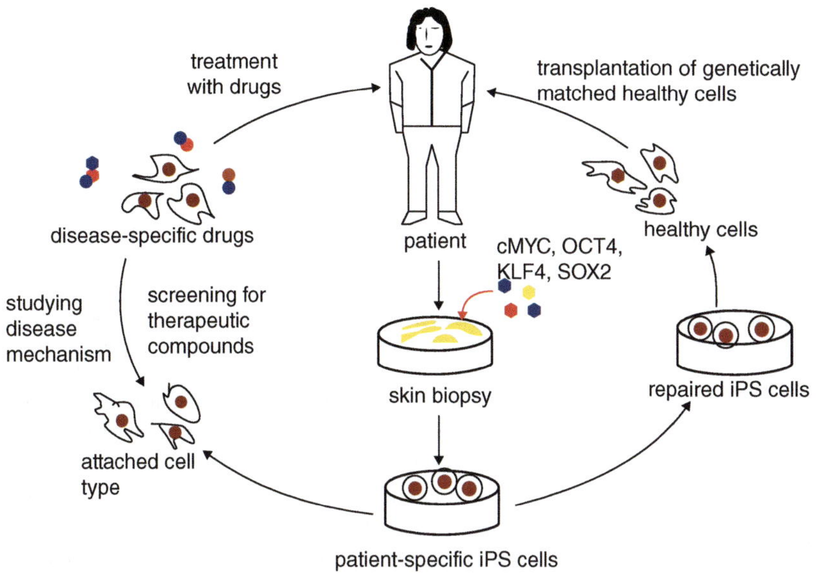

FIGURE 5.10 Use of iPS cells for drug discovery and for analysis and treatment of genetic diseases [1]:1259. The left side of the diagram shows how differentiated cells that are generated from iPS cells derived from a patient with a genetic disease can be used for analysis of the disease mechanism and for discovery of therapeutic drugs. The right side of the diagram shows how the genetic defect might be repaired in the iPS cells, which could then be induced to differentiate in an appropriate way and grafted back into the patient without danger of immune rejection.

Tissue repair by transplantation, however, is not the only application for which ES, iPS, and transdifferentiated cells can be used: there are often ways in which they promise to be more immediately valuable. In particular, they can be used to generate large, homogeneous populations of specialized cells of any chosen type in culture; and these can serve for investigation of disease mechanisms and in the search for new drugs acting on a specific cell type (Figure 5.10) [1]:1259. That is, iPS cells and transdifferentiated cells can be used for investigation of disease mechanisms and in search for new drug discoveries.

There is a problem when a disease has a genetic cause. In order to solve the problem, we can derive iPS cells from sufferers and use these cells to produce the specific cell types that malfunction, to investigate how the malfunction occurs, and to screen for drugs that might help to put it right. Timothy syndrome provides an example. In this rare genetic condition, there is a severe life-threatening disorder in the rhythm of the heartbeat (as well as several other abnormalities), as a result of a mutation in a specific type of Ca^{2+} channel. To study the underlying pathology, researchers took skin fibroblasts from patients with the disorder, generated iPS cells from the fibroblasts, and drove the iPS cells to differentiate into heart muscle cells. These cells, when compared with heart muscle cells prepared similarly from normal control individuals, showed irregular contractions and abnormal patterns of Ca^{2+} influx and electrical activity that could be characterized in detail. From this finding, it is a small step to the development of an in vitro assay for drugs that might correct the misbehavior of the heart muscle cells.

This approach to drug discovery whereby iPS cells are prepared from the individual patient, differentiated into the relevant cell type, and used to test candidate drugs in vitro would seem to represent a huge advance on the slow, costly traditional methods that involve administration of test compounds to large numbers of people.

5.11 ENHANCING THE IMMUNE RESPONSE AGAINST THE SPECIFIC TUMOR

In this section, we describe the immunological therapies with antibodies targeting tumor cells, and those with specific antibodies preventing tumor cells from engaging with the inhibitory receptors on T-cells [1]:1137. For more than a century it has been a dream of cancer researchers to somehow harness the immune system in a controlled and efficient way to exterminate cancer cells, just as it exterminates infectious organisms. There are, finally, signs that this dream may one day be realized, at least for some forms of cancer.

The simplest type of immunological therapy, conceptually at least, is to inject the patient with antibodies that target the cancer cells. This approach has had some successes. About 25% of breast cancers, for example, express unusually high levels of the HER2 protein. A monoclonal antibody called trastuzumab (trade name Herceptin) that binds to HER2 and inhibits its function, slows the growth of breast tumors in humans that overexpress HER2, and it is now an approved therapy for these cancers. A related approach uses antibodies to deliver poisons to the cancer cells. Antibodies against proteins that are abundant on the surface of a particular type of cancer cell but rare on normal cells can be armed with a toxin that kills those cells that bind the antibody molecule.

A great deal of current excitement centers on a different type of approach, based on the relatively recent recognition that the microenvironment in a tumor is highly immunosuppressive. As a result, the immune system victimized by cancer is prevented from destroying the tumor cells. From the thousands of genome sequences thus far determined, we know that a typical cancer cell will contain on the order of 50 proteins with a mutation that alters an amino acid sequence, most of these being passenger mutations. Many of these mutant proteins will be recognized by the patient's immune system as foreign, but – to allow the cancer cells to survive throughout the course of tumor progression – the cancer cells have evolved a set of anti-immune defense. These defenses include the expression on the cancer cell surface of one or more proteins that bind to inhibitory receptors on activated T-cells.

The normal immune system is subject to complex controls that keep its activity within safe bounds and prevent autoimmunity from developing. The inhibitory receptors that are expressed on the surface of activated T-cells have an important normal function: they control the immune response by down-regulating the T-cell response under appropriate circumstances, preventing autoimmunity from developing.

However, in the context of a tumor, the down-regulation is inappropriate, because it prevents the T-cell from killing the cancer cells that are threatening its survival.

In its attack on infectious organisms, the natural immune system usually eliminates every last trace of infection and maintains this immunity in the long term. The challenge is to find ways of recruiting the immune system to attack cancers with similar efficiency and specificity, hunting the cancer cells down by virtue of the tumor-specific antigens that they express. With this aim, a new type of anti-cancer therapy focuses on overcoming the immunosuppressive environment in a tumor through the use of specific antibodies that prevent the tumor cells from engaging with the inhibitory receptors on T-cells. Blocking the action of the immune suppressors with such treatments should unleash an immune attack on the cancer cells (see Chapter 7). That is, the inhibitory receptor on the surface of T-cells have the important normal function of preventing autoimmunity. However, the tumor cells use the inhibitory receptor for their survival. A new type of anti-cancer therapy uses specific antibodies that prevent the tumor cells from engaging with the inhibitory receptor on T-cells.

E5.1 RAS ONCOGENES

In the early 1980s, it was discovered that cancers can be caused by mutations from a normal human gene, one of a small family of proto-oncogenes called "Ras." This implies that cancers are caused by mutations in a limited number of cancer-critical genes.

Normal Ras proteins are monomeric GTPases that help to transmit signals from cell-surface receptors to the cell interior. The Ras oncogenes isolated from human tumors contain point mutations that create a hyperactive Ras protein. The Ras protein plays an important role in cell signaling. In its GTP-bound form, it is active and stimulates a cascade of protein phosphorylations in the cell. Most of the time, however, the protein is in its inactive, GDP-bound form. It becomes active when it exchanges GDP for a GTP molecule in response to extracellular signals, such as growth factors, that bind to receptors in the plasma membrane [1]:157, 855. GTP (guanosine triphosphate [1]:85, 101) becomes GDP (guanosine diphosphate) by hydrolyzing. Due to mutation, Ras protein cannot shut off the hyperactivity, causing cancer. One or another of the three human Ras family members is mutated in perhaps 30% of all human cancers. Ras genes are among the most important of all cancer-critical genes [1]:156–157, 1106.

E5.2 STEPS OF TUMOR PROGRESSION

Mutations inactivating the tumor-suppressor gene Apc (adenomatous polyposis coli) gene, in most cases, appear to be the first. Changes that lead to genetic instability of the Apc gene are likely to arise early in tumor progression.

Activating mutations in the proto-oncogene K-Ras (a member of the Ras gene family) occur later. Inactivating mutations in p53 are thought to come later still. Loss of p53 function allows cancer cells to endure stress, avoiding apoptosis and cell cycle arrest. Additionally, loss of p53 is related to the heightened activation of oncogenes such as Ras. Only by losing p53 function can the cancer cells with hyperactive oncogenes survive and progress [1]:1123–1126.

E5.3 CANCER AS DISTURBANCES

The cells of a multicellular organism coordinate their behavior, sending, receiving, and interpreting an elaborate set of extracellular signals that serve as social controls. As a result, each cell behaves in a social manner – resting, growing, dividing, differentiating, or dying – as needed for the organism.

Disturbances that upset that harmony mean trouble for a multicellular organism. In a human body with more than 10^{13} (ten trillion) cells, billions of cells experience mutations every day, potentially disrupting the social controls. Most dangerously, a mutation may give one cell a selective advantage, allowing it to grow and divide slightly more vigorously and survive more readily than its neighbors and in this way to become a founder of a growing mutant clone. A mutation that promotes such selfish behavior by individual members can jeopardize the future of the whole organism. Over time, repeated rounds of mutation can cause matters to go from bad to worse. These are the basic ingredients of cancer, which is a disease with an individual mutant clone of cells prospering at the expense of its neighbors. In the end – as the clone grows, evolves, and spreads – it can destroy the entire cellular society [1]:1091. That is, each cell of a multicellular organism behaves in a social manner as needed for the organism. Disturbance upsetting the harmony occurs by a mutation that may give one cell a selective advantage, allowing it to proliferate slightly more vigorously and survive more readily than its neighbors. Repeated rounds of mutation can cause the matter to go from bad to worse, destroying the entire cellular society.

E5.4 CT (COMPUTED TOMOGRAPHY)

An X-ray photograph obtains only one directional information about the human body. Contrary to X-rays, the CT can provide all directional information of an X-ray projected image. The apparatus was invented by Godfrey N. Hounsfield and Allan M. Cormack. They were awarded the Nobel Prize in Physiology and Medicine in 1979.

Using the mathematical tool Fourier Transform (the Fast Fourier Transform [FFT] was used for the purpose of fast computing, in fact), they effectively obtained all the directional information [9]. The CT can detect 0.5 mm size and hence the apparatus can be used for clinical detection of a tumor in initial stage of its early growth.

E5.5 P. EHRLICH

For the first time, Paul Ehrlich used the concept of chemotherapy. From 1878 to 1883, he studied internal medicine at the University of Berlin. In 1892, he discovered immunity in mother's milk. He established the theory that probed the relation between antibodies and antigens. He founded the private Institute for Serum Research and Serum Testing. He discovered the drug against Trypanosoma in 1904,

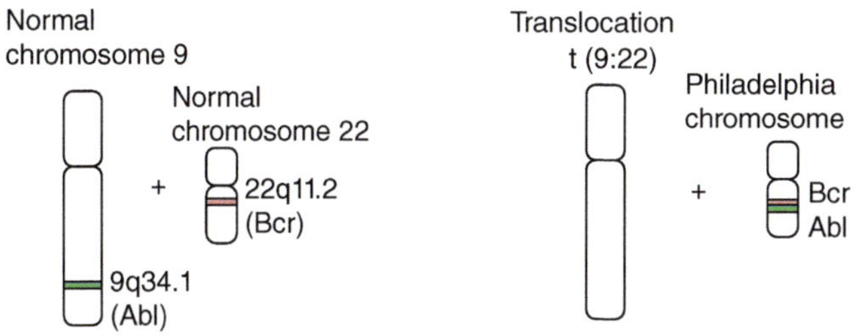

FIGURE E5.1 Schematic of the Philadelphia chromosome formation [10].

and the drug against sleeping sickness in 1906. Thus, he researched chemotherapy. Kiyoshi Shiga, who discovered dysentery bacillus, was a student of the same professor. P. Ehrlich was awarded the Nobel Prize in Physiology and Medicine together with E. Metchnikoff in 1908.

E5.6 ENZYME IDO (INDOLEAMINE 2,3-DIOXYGENASE)

The enzyme IDO (indoleamine 2,3-dioxygenase) catabolizes tryptophan, an essential amino acid, in order to produce the immunosuppressive metabolite kynurenine (Figure 5.3). IDO seems to function in maintaining balance between the immune response and tolerance during infections, but can be induced to the escape phase during the equilibrium phase of cancer development. Finally, cancer can produce materials such as collagen, creating a physical barrier to interaction with cells of the immune system [3]:719. That is, IDO can be induced to the escape phase of cancer, which finally can produce materials such as collagen, creating a barrier to interaction with immune system.

E5.7 THE PHILADELPHIA CHROMOSOME

The chromosomal defect in the Philadelphia chromosome is a reciprocal translocation, in which parts of two chromosomes, 9 and 22, swap places, resulting in a fusion gene created by juxtaposing the Abl gene on chromosome 9 (region q 34) to a part of the Bcr (breakpoint cluster region) gene on chromosome 22 (region q 11). This chromosomal translocation is designated as t (9:22) (q34:q11) (Figure E5.1).

The symbol Abl is derived from Abelson, the name of a leukemia virus. The symbol Bcr is derived from breakpoint cluster region. The translocation results in an oncogenic Bcr-Abl gene fusion that can be found on the shorter derivative chromosome 22 (Figure E5.1). This gene encodes for a Bcr-Abl fusion protein.

Three clinically important variants encoded by the fusion gene are the p190, p210, and p230 isoforms. p190 is associated with B-cell acute lymphoblastic leukemia, while p210 is associated with CML (section 5.9). p230 is associated with chronic myelogenous leukemia associated with neutrophilia and thrombocytosis (CML-N) [10].

6 NIR Photoimmunotherapy

Anti-cancer therapies consist of surgery, radiation therapy, and chemotherapy. In all these therapies, the treatment damages not only cancer cells but also neighboring normal cells. Thus, conventional therapies have burdened patients with side effects.

Based on the theory unifying chemistry, physics, and biology, a new therapy specifically removing cancer cells without damaging normal cells appeared. It was developed using the near-infrared (NIR) light by Hisataka Kobayashi (E6.1) at the National Institutes of Health (NIH). In conventional therapy, there is the method using antibody-drug conjugate (ADC) to specifically target and kill cancer cells. However, this method has the severe problem of damaging neighboring normal cells.

6.1 NEW FORM OF ADC (ANTIBODY-DRUG CONJUGATE)

In order to solve the problem just mentioned, H. Kobayashi developed a new form of ADC; that is to say, he attached a photoabsorbing chemical, termed IRDye 700DX (IR700), to the antibody. The conjugate is injected into the body, and the target cancer tissue is illuminated by the near-infrared light with a wavelength of 690 nm. The IR700 part of the conjugate is activated, splitting and turning hydrophobic, leading to damaging cell membranes and killing cancer cells (Figure E6.1). In this method, only target cells binding IR700 are illuminated and killed (Figures 6.1, 6.2). Therefore, this method is safer than conventional ADC.

In order to propagate sufficient light for therapy into depth, it is necessary to use near-infrared light with a wavelength of 700 nm to 850 nm to avoid visible light, highly absorbed by hemoglobin; light with a wavelength of 690 nm is used now because 690 nm penetrates skin and tissue to several centimeters deep without damaging any normal cells.

This new method was termed near-infrared photoimmunotherapy (NIR-PIT). Cancer-specific targets are broadened by changing the antibody, targeting antigens such as proteins EGFR (Figure 5.7), HER2, PSMA, CD25, CEA, mesothelin, GPC3, CD20, and PD-L1. Because NIR-PIT can achieve spatially selective removal of target cells, it can be used for eliminating cancer stem cells such as breast cancer and glioblastoma stem cells with cancer stem markers of CD44 and CD133. The therapy NIR-PIT targeting cancer stem cells suppresses regrowth of cancer for a long period [11].

6.2 TARGETING SYSTEMIC METASTASIS

NIR-PIT has a strong prospect in indirect cancer immunotherapy, achieving spatially selective depletion of cancer-associated immunosuppressing regulatory T-cells (T_{reg}).

DOI: 10.1201/9781003780663-6

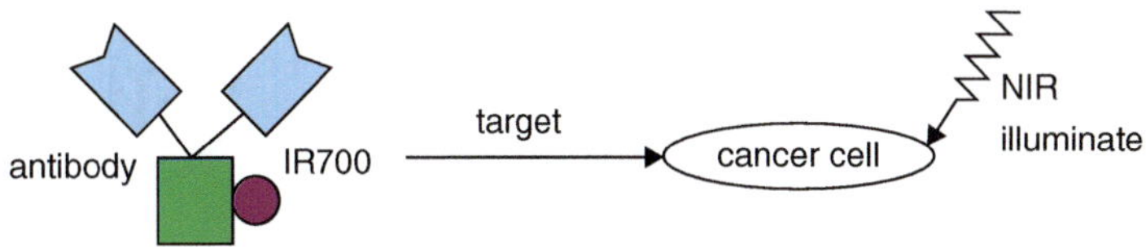

FIGURE 6.1 The process of NIR-PIT.

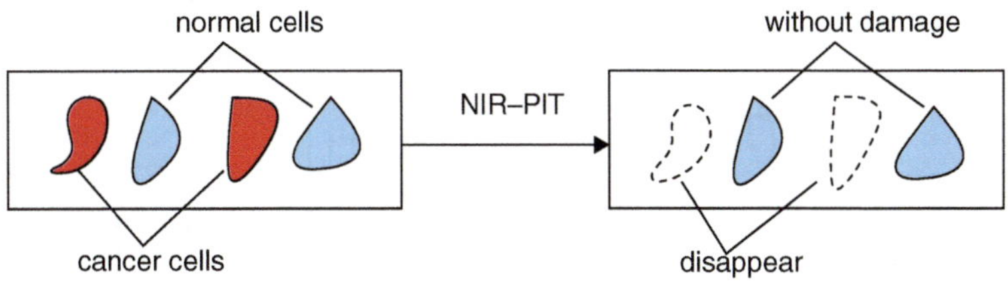

FIGURE 6.2 Normal cells in the neighborhood of cancer cells do not receive damage in NIR-PIT.

T_{reg} inhibits anti-cancer attack by CTL. The elimination of *Treg* near cancer allows neighboring CTL to attack cancer within 1 hour (Figure 6.3). Remarkably, T_{reg}-targeting NIR-PIT causes selective systemic regression of untreated distant metastatic cancer with the same cell origin as treated cancer. The reason is that presumably CTL once activated is no longer susceptible to inactivity induced by T_{reg}. Untreated distant metastatic cancer (E6.4) of the same type as treated cancer cells is killed thus by NIR-PIT targeting T_{reg}. That is to say, untreated distant metastatic cancer cells disappear by treating only original cancer cells (Figure 6.4).

6.3 CLINICAL TRIAL

The patent was authorized to Aspyrian Therapeutics Inc. (since renamed Rakuten Medical, Inc.). In June 2015, a phase I clinical trial was initiated in the US to treat patients suffering from head and neck cancer with drug the cetuximab-IR700 conjugate (RM-1929). In the US, phase II ended in 2018, and the FDA planned to initiate phase III. Head and neck cancer represents antigen EGFR (Figure 5.7) on its surface. Similar clinical trials are planned for lung, esophageal, bladder, and pancreatic cancers.

6.4 IMMUNITY DUE TO T-CELLS

In NIR-PIT, various antigens are released from dying cancer cells illuminated by the near-infrared light. The antigens are digested by dendritic cells, which are innate immune cells, expressing MHC class I-peptide complex on the surface of dendritic cells and activating T-cell. Activated T-cells acquire immunological memory and have the function of a vaccine (Figure 6.5). Therefore, in a mouse completely cured

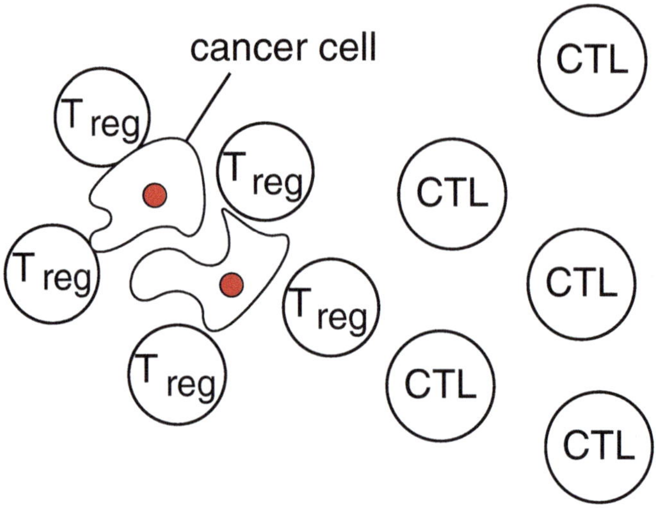

FIGURE 6.3 CTL can attack cancer by eliminating T_{reg}.

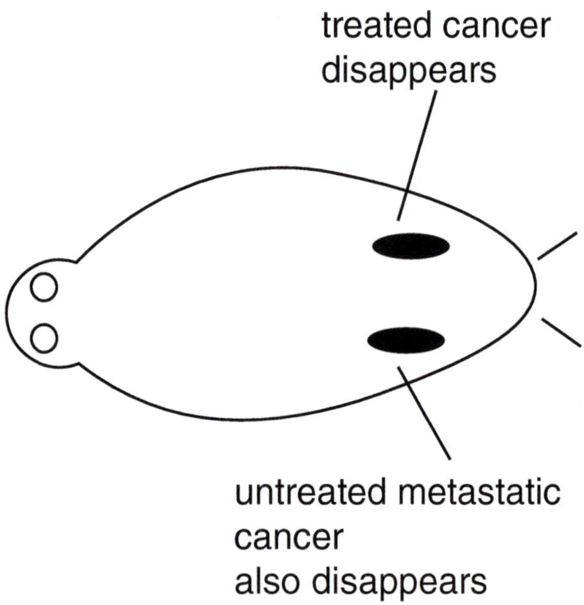

FIGURE 6.4 Untreated metastatic cancer disappears by NIR-PIT.

by NIR-PIT, cancers did not relapse even when the same cancer cells were introduced. This is why this therapy is referred to as photoimmunotherapy.

Most cancer cells do not express MHC class I-peptide complex and so they escape from the immune response. In NIR-PIT, antigens released from dying cancer cells, however,

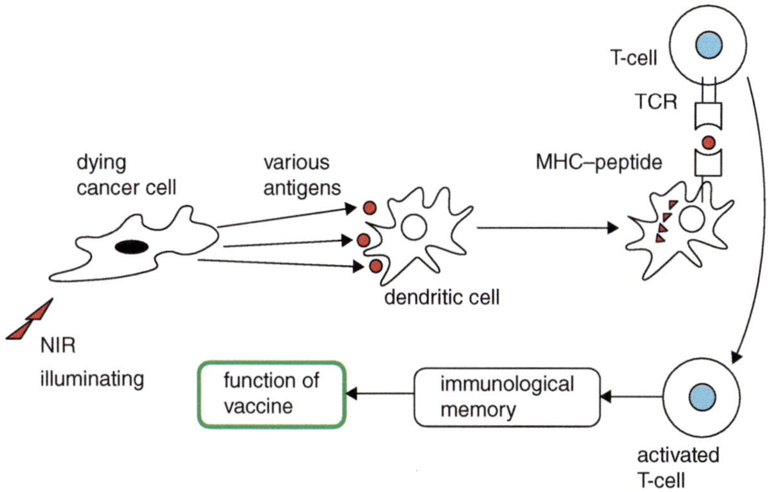

FIGURE 6.5 Antigens released from dying cancer cells activate T-cells, leading to a vaccine's function [12].

are digested by dendritic cells, which express MHC class I-peptide complex on the surface. As a result, the peptide of the antigen is not hidden by the cancer cell, but expressed by dendritic cells certainly, resulting in the function of immunity. By using antigen-specific antibodies, theoretically 80–90% of cancer cells can be treated by NIR-PIT.

6.5 NEAR-INFRARED PHOTOIMMUNOTHERAPY FOR VARIOUS CANCERS

In this section, we discuss a range of NIR-PIT applications for various cancers in clinical and preclinical studies [13].

6.5.1 Head and Neck Squamous Cell Carcinoma (HNSCC)

HNSCC is the seventh most common cancer with more than 931,000 new cases worldwide, and is the seventh most common cause of cancer death with more than 467,000 deaths worldwide in 2020. Cetuximab-IR700 was the first agent clinically introduced for NIR-PIT in 2006. Phase I and II clinical trials of NIR-PIT using cetuximab-IR700 in patients with recurrent HNSCC concluded in 2018 and showed that cetuximab-IR700 NIR-PIT was more effective than current second- and third-line therapies for recurrent HNSCCs. These results prompted the FDA to assign a fast-track designation for cetuximab-IR700 in 2018. A global phase III trial in recurrent HNSCC is currently underway.

6.5.2 Glioblastoma Multiforme (GBM)

Burley et al. reported that EGFR-targeted NIR-PIT showed effectiveness in a xenograft model of GBM. Therefore, EGFR- or CSC (cancer stem cell)-targeted NIR-PIT is a potential for GBM.

6.5.3 Esophageal Cancer

Esophageal carcinoma is the eleventh most common cancer with more than 604,000 cases worldwide and is the sixth most common cause of cancer death with more than 544,000 deaths worldwide in 2020. The two major subtypes of esophageal cancer are esophageal squamous cell carcinoma (ESCC) and esophageal adenocarcinoma (EAC). ESCC represents 90% of all cases of esophageal cancer globally and is dominant in East Asia, East Africa, and South America. EAC is more common in developed countries than in developing countries.

In in vitro studies, EGFR- or HER2-targeted NIR-PIT has been shown to be effective in esophageal carcinoma on all lines. Many studies have emphasized the importance of cancer-associated fibroblasts (CAFs) in esophageal carcinoma. CAF-targeted NIR-PIT inhibited tumor progression in co-culture models of ESCCs and CAFs. Therefore, esophageal cancer could be treated with some combination of EGFR-, HER2-, or CAF-targeted NIR-PIT.

6.5.4 Lung Cancer

Lung cancer is the second most common cancer with more than 2,206,000 new cases worldwide and is the most common cause of cancer death with more than 1,796,000 deaths worldwide in 2020. The two major subtypes of lung cancer are non-small cell lung cancer (NSCLC) (85% of patients) and small cell lung cancer (SCLC) (15%).

EGFR overexpression has been identified in 40–80% of NSCLC. NIR-PIT with panitumumab (another antibody targeting EGFR)-IR700 conjugates inhibited tumor growth in a transgenic mouse model of spontaneous lung cancer expressing human EGFR. Several monoclonal antibodies (mAbs) against PD-L1 and PD-1 have demonstrated clinical benefit in patients with NSCLC and are collectively known as checkpoint inhibitors. PD-L1 is overexpressed in many cancers, and therefore, it is a potential target for NIR-PIT. NIR-PIT using avelumab (human anti-PDL1 mAb)-IR700 induced significant therapeutic effects in an NSCLC xenograft model. Thus, PD-L1 targeted NIR-PIT might be useful for NSCLC cancers with high PD-L1 expression.

In xenograft models, HER2-targeted NIR-PIT led to significant reduction in pleural dissemination by HER2-expressing NSCLC cells.

The optical fibers are utilized through the bronchoscope to deliver NIR light to key areas. This bronchoscope technique can be navigated with a 3D navigation system from CT images beforehand to guide the fiber precisely to the tumor in the lung. By combining this with an endoscope or catheter, it is possible to deliver light locally within the body.

6.5.5 Breast Cancer

Female breast cancer is the most common cancer with more than 2,260,000 new cases worldwide and is the fifth most common cause of cancer death with more than 684,000 deaths worldwide in 2020. Tumors are classified by their HR and HER2 status: HR+/HER2– (70% of patients), HR+/HER2+ (15–20%), and triple-negative (HR-/HER2-; 10–15%).

Trastuzumab, an anti-HER2 mAb, has been used as a NIR-PIT agent in xenograft models of breast cancer. For triple-negative breast tumors, no targeted therapy is currently available. However, EGFR expression has been reported in 50–89% of cases. Therefore, EGFR-targeted NIR-PIT could be used in some cases of triple-negative breast cancer. In xenograft models with two different cell lines established from triple-negative breast cancer, cetuximab-IR700 NIR-PIT suppressed tumor growth and prolonged survival.

6.5.6 Liver Cancer

Liver cancer is the eighth most common with more than 905,000 new cases worldwide and is the third common cause of cancer death with more than 830,000 deaths worldwide in 2020. Hepatocellular carcinoma (HCC) is the most common type of primary hepatic malignancy. HCC is caused by chronic viral hepatitis which results in cirrhosis and tumor formation. In general, mortality rates are high for HCC.

Glypican-3 (GPC3) is highly expressed in HCC, and is a target candidate for NIR-PIT. GPC3-targeted NIR-PIT inhibited tumor growth compared to untreated controls in a xenograft model of HCC. Newly developed catheter-based devices containing LEDs could be used to deliver light for NIR-PIT resulting in tumor suppression in xenograft models.

6.5.7 Pancreatic Cancer

Pancreatic cancer is the fourteenth most common cancer with more than 495,000 cases worldwide and the eighth most common cause of cancer death with more than 466,000 deaths worldwide in 2020.

The pancreas primary is difficult to approach with light. It is possible that endoscopic ultrasound light delivery could provide light to the tumor.

Cadherin-17 (CDH17) is highly expressed on gastrointestinal cancer cells. CDH17-targeted NIR-PIT inhibited tumor growth in a xenograft model which used a pancreatic cancer cell line. TROP2 is also overexpressed in pancreatic cancer. TROP2-targeted NIR-PIT inhibited tumor growth in pancreatic cancer xenograft models.

E6.1 H. KOBAYASHI

Hisataka Kobayashi was born in Nishinomiya, Japan in 1961. In 1987, he graduated from the Department of Medicine at Kyoto University, where in 1995, he also received a doctorate degree. He was inaugurated as a US NCI (National Cancer Institute)/NIH senior fellow in 1995. In 2004, as the NCI molecular imaging program's chief researcher, he developed NIR-PIT.

NIR-PIT results in the function of a vaccine. That is to say, various antigens released from cancer cells destroyed by NIR-PIT are sensed by dendritic cells that activate T-cells to have immunological memory, resulting in the function of a vaccine for cancer antigens [12].

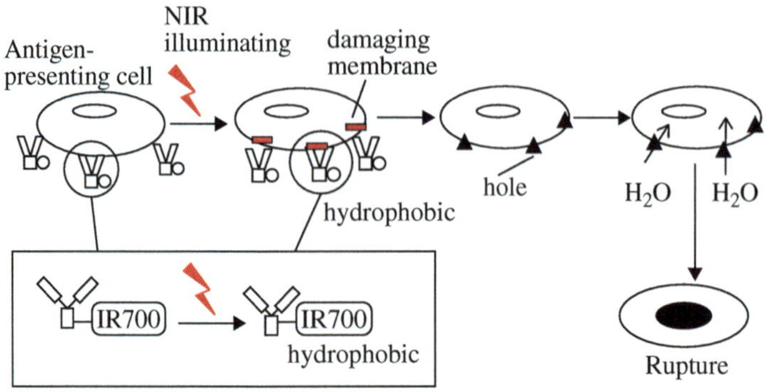

FIGURE E6.1 The mechanism of cancer rupture by NIR-PIT [14].

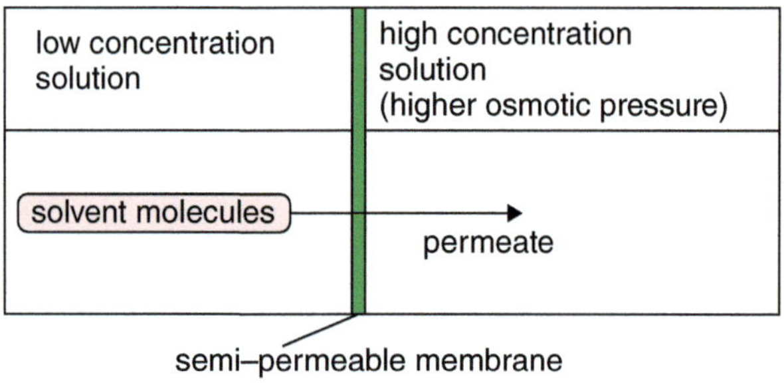

FIGURE E6.2 Solvent molecules pass through a semi-permeable membrane to a higher-concentration solution.

E6.2 MECHANISM OF CANCER RUPTURE

Firstly, antibody-IR700 conjugate binds on an antigen-presenting cell's surface. Once the NIR light irradiates, the conjugate IR700 differentiates in characteristic from water-soluble to hydrophobic. The photochemical reaction causes antibody-IR700 to aggregate on the antigen-presenting cell's membrane, damaging the membrane and making holes on the surface, which allow water to flow into the antigen-presenting cell whose osmotic pressure (E6.3) is higher than that outside the cell. As a result, the ultimate high pressure within the antigen-presenting cell ruptures the cell, leading to cell death (Figure E6.1) [14].

E6.3 OSMOTIC PRESSURE

Osmotic pressure is the minimum pressure which needs to be applied to a higher-concentration solution to prevent the inward flow of its pure solvent across a semi-permeable membrane. Osmotic pressure is proportional to the concentration of solute.

When two solutions containing different concentrations of solute are separated by a semi-permeable membrane, the solvent molecules pass through the semi-permeable membrane from the low-concentration solution to the higher-concentration solution where osmotic pressure is higher (Figure E6.2). For example, when we wash our eyes with pure water, we feel pain. The water molecules permeate into our eyes whose osmotic pressure is higher than that of the pure water, and the eyes expand. This is why we feel pain.

E6.4 METASTASIS

As long as the neoplastic cells have not yet become invasive, the tumor is said to be benign. For most types of such neoplasms, removing or destroying the mass locally usually achieves a complete cure.

A tumor is considered a true cancer if it is malignant; that is to say, when its cells have acquired the ability to invade surrounding tissue. Invasiveness is an essential characteristic of cancer cells. It allows them to break loose, enter blood or lymphatic vessels, and form secondary cancers called metastasis at other sites in the body. In general, the more widely a cancer spreads, the harder it becomes to eradicate. It is generally metastasis that kills the cancer patient.

E6.5 $CD4^+$ REGULATORY T-CELLS

Following several findings in the 1970s and 1980s, suggesting the existence of T-cells capable of suppressing experimental autoimmune diseases in rodents, efforts were made to characterize such auto-immunosuppressive $CD4^+$ T-cells (T_{reg}) by reliable molecular markers. These efforts culminated in the discovery in the mid-1990s that T_{reg} constitutively and highly expressed CD25 (IL-2 receptor α chain). Removal of $CD25+CD4^+$ T-cells, which constituted 10% of peripheral $CD4+$ T-cells, indeed produced a wide spectrum of autoimmune diseases immunologically similar to human counterparts such as autoimmune gastritis, thyroiditis, and type 1 diabetes in otherwise normal mice; reconstruction of $CD25^+CD4^+$ T-cells inhibited disease development. The phenotypic characterization has facilitated the elucidation of their function in vivo and in vitro and their roles in antitumor, antimicrobial, and transplantation immunity in animals, and enabled in the early 2000s the identification of human $CD25^+CD4^+$ T_{reg}s with an equivalent function and phenotype to those in mice [15].

Mary E. Brunkow, Fred Ramsdell, and Shimon Sakaguchi, who had researched T_{reg}, were awarded the Nobel Prize in Physiology and Medicine in 2025.

7 Checkpoint Blockade

7.1 CANCER IMMUNOTHERAPY

To strengthen the natural immune response to cancer is the approach of cancer immunotherapy. In the approach, there are two methods, one of which is to induce expression of a co-stimulatory molecule such as B7 on the cancer surface, binding B7 on a cancer cell's surface to CD28 on a T-cell's surface leading to activation of naive T-cells to effector T-cells proliferating as shown in Figure 7.1. But CTLA-4 has a higher affinity than CD28 for B7, providing one mechanism by which it could block the co-stimulation of T-cells (Figure 7.2) [3]:287. Similarly, cancer cells may be transfected with genes encoding GM-CSF (granulocyte-macrophage colony-stimulating factor) in order to induce the maturation of cancer-proximal monocytes into monocyte-derived dendritic cells. Once these cells have differentiated and capture antigens from the cancer, they may migrate to local lymph nodes and activate cancer-specific T-cells. The efficacy in this therapy, however, was not approved [3]:727. The monocytes receiving signals from chemokines migrate across the vascular endothelium into infected and inflamed tissues, and are also able to differentiate into dendritic cells, depending on the signals that they receive from their environment [3]:86. IL-2 produced by T_{H1} cells augments effector T-cell differentiation. T_{H1} cells also produce other cytokines such as IL3 and GM-CSF which stimulate the production of new monocytes by acting on hematopoietic stem cells in the bone marrow [3]:460.

Another method of cancer immunotherapy is the checkpoint blockade that interferes with normal inhibitory signals regulating lymphocytes. The immune responses are controlled by several positive and negative immunological checkpoints. Negative immunological checkpoints are provided by inhibitory receptors such as CTLA-4 and PD-1 [3]:727.

7.2 CANCER'S ESCAPE FROM IMMUNITY

Although T-cells attack antigen-presenting cells (APC), excessive activated T-cells may also attack normal cells that are not foreign antigens, inducing an autoimmune response. Receptor CTLA-4 on a T-cell's surface plays a role in preventing an autoimmune response. That is to say, as shown in Figure 7.2 left, binding B7 on a dendritic cell's surface to checkpoint CTLA-4 on a T-cell's surface releases a signal inhibiting the T-cell, preventing an autoimmune response. In a normal state, a checkpoint is used for preventing difficulties such as autoimmune responses. Cancer cells, however, use the checkpoint to escape the immune system and survive. That is to say, binding B7 on a cancer cell's surface to CTLA-4 on a T-cell's surface leads to inhibition of T-cell activity, stopping the T-cell's attack on cancer and helping the cancer to escape from the immune response and survive.

DOI: 10.1201/9781003780663-7

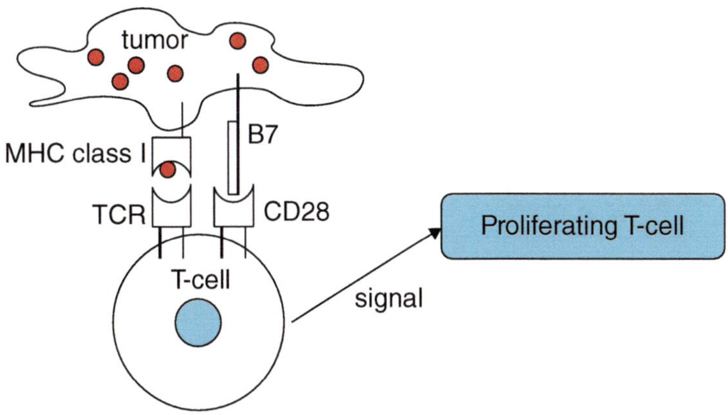

FIGURE 7.1 Inducing B7 results in activating naive T-cells [3]:371.

7.3 IMMUNE CHECKPOINT BLOCKADE (ICB)

To prevent cancer from misusing checkpoints, J.P. Allison blocked CTLA-4 on a T-cell's surface with anti-CTLA-4 antibody, termed the checkpoint blockade, in order to keep the T-cell's attack contained to cancer (Figure 7.2 right). Checkpoint blockade therapy based on anti-CTLA-4 antibody ipilimumab was found to be effective in treating metastatic melanoma and received FDA approval. Patients treated with ipilimumab showed increased activity of T-cells recognizing cancer-testis antigen NY-ESO-1 expressed on melanoma (E7.3). As a side effect of ipilimumab, autoimmune phenomena seemed to be increased [3]:728.

Another checkpoint is the inhibitory receptor PD-1 on a T-cell surface. Binding PD-L1 on a cancer cell surface to PD-1 leads to inhibition of the T-cell's attack on cancer (Figure 7.3). A checkpoint blockade with anti-PD-1 antibody nivolumab increases the T-cell's activity to cancer. For metastatic melanoma, nivolumab received FDA approval.

In 1992, Yasumasa Ishida, who was a graduate student of Professor T. Honjo's laboratory at Kyoto University, discovered a molecule on the T-cell's surface whose expression was augmented upon inducing the T-cell's death. He named the molecule "PD-1 (programmed death-1)." T. Honjo proved that PD-L1 on the surface of an antigen-presenting cell (APC) binds to PD-1 and inhibits the function of the immune response. He also proved that most cancer cells express PD-L1, probing the escape of cancer from the immune system. T. Honjo reached the conclusion that blocking the inhibitory receptor PD-1 leads to preventing the escape of cancer from the immune response. On the basis of this concept, he developed immune checkpoint blockade therapy. Theoretically, the therapy is effective on every type of cancer cell, because it is a method that augments the natural immune system. Checkpoint blockade immunotherapy provides the fourth cancer therapy. Drugs targeting PD-1 still present drawbacks: more than half of patients do not respond to this therapy [16].

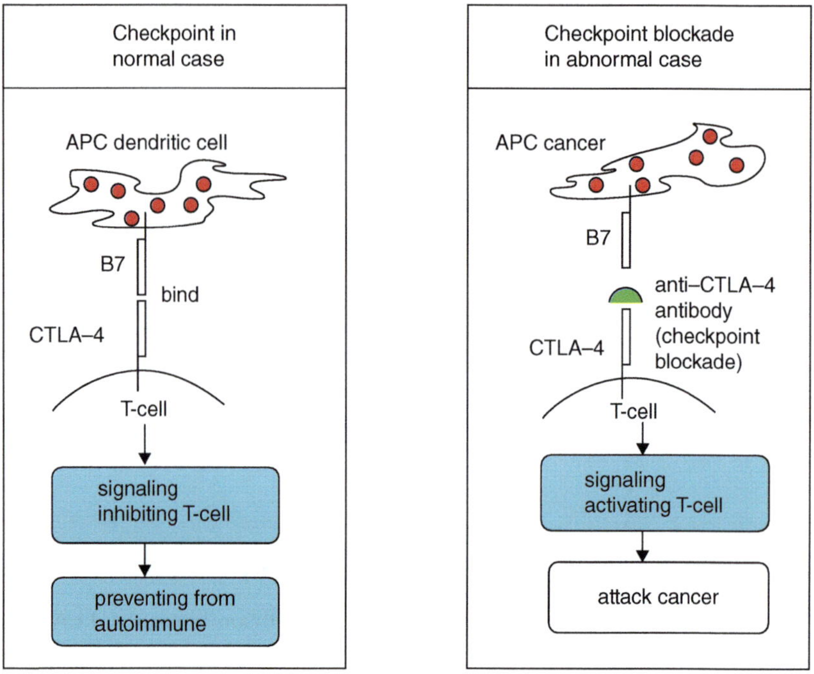

FIGURE 7.2 Checkpoints in normal case and abnormal case.

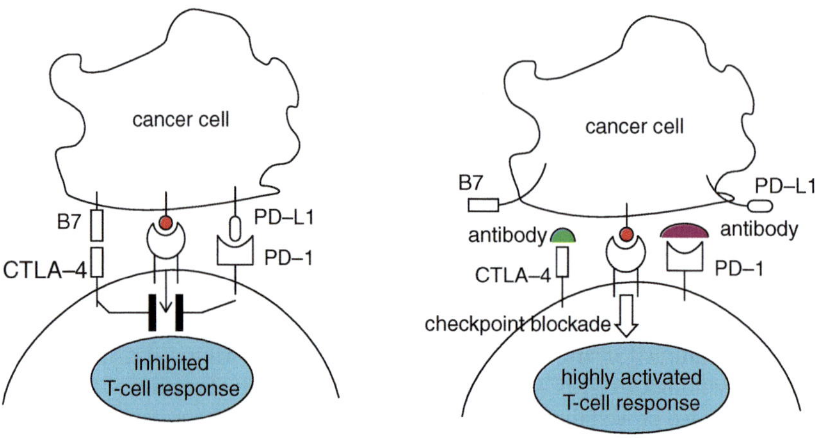

FIGURE 7.3 Cancer therapy by immune checkpoint blockade.

7.4 CLINICAL RESPONSE TO ICB THERAPY

One of the hallmarks of immunotherapy is the durability of the responses that can be translated into survival benefit. Only a subset of tumor histology and a small percentage of the patients in each histology, however, are responsive to these checkpoint blockade therapies. The response rates of different tumor types to PD-1/PD-L1 checkpoint blockade tend to be proportional to their corresponding "tumor mutational burden (TMB)" (E7.4), presumably from the immunogenic neoantigens that are recognized as foreign by cytotoxic T-lymphocytes (CTLs). The response to immune checkpoint blockade therapy, however, is complex, heterogeneous, and inconsistent, and additional mechanisms are at play. Increased PD-L1 expression has been correlated with the immune response and is currently used as a biomarker for ICB therapy in NSCLC (non-small cell lung cancer) and urothelial carcinoma. Moreover, elevated numbers of tumor-infiltrating lymphocytes (TILs) have been noted in responsive cancers [17].

The ability of tumors to induce an adaptive immune response relies on recognition of cancer cells as foreign. High TMB with accompanying elevated neoantigen expression plays an important role in antitumor immunity. Van Allen et al. demonstrated that mutational load was significantly associated with the response to anti-CTLA-4 treatment in patients with metastatic melanoma. Additionally, Rizvi et al. noted that the response to anti-PD-1 treatment correlated with high TMB and neoantigen load in patients with NSCLC. In keeping with these studies, immunogenic tumors with low TMB, such as pancreatic and prostate cancers, are inherently more resistant to treatment with checkpoint inhibition. Extrapolating from these data, mechanisms leading to loss of neoantigen expression by cancer cells may result in acquiring resistance to ICB therapy.

The presence of PD-L1-expressing cancer within a tumor is known to be an important predictor of response to ICB therapy and is commonly used as a biomarker. It has been shown that tumors lacking PD-L1 expression generally show inferior clinical response to ICB compared with those with higher levels of ligand. Regardless of PD-L1 expression, tissues that lack TILs are unlikely to respond to ICB therapy. Tumors with larger numbers of TILs demonstrate greater response to ICB therapy and TILs may serve as another predictive "biomarker." A study in patients with metastatic melanoma showed that pre-existing tumoral CTLs are a qualification for the response to anti-PD-1 therapy.

Abnormalities in the MAPK (mitogen-activated protein kinase) pathway promote oncogenesis in multiple tumor types. Increased MAPK signaling impairs recruitment and function of TILs through expression of VEGF (Figure 5.7) and multiple inhibitory cytokines, and leads to immune evasion. WNT/β-catenin signaling is another canonical oncogenic pathway involved in many essential cellular processes. Abnormalities in WNT/β-catenin signaling have been observed in multiple tumor histologies and are correlated with increased cell invasiveness and metastatic potential. Constitutive WNT signaling with stabilization of β-catenin, in addition to oncogenesis, might be involved in the ICB resistance through tumor T-cell exclusion.

Vascular endothelial growth factor (VEGF) functions as an immunosuppressive cytokine and is associated with the resistance to ICB. VEGF levels were found to be higher in anti-PD-1 therapy nonresponders compared with responders. Furthermore, VEGF increases expression of inhibitory receptors, contributing to CTL exhaustion. Corroborating this evidence, inhibition of VEGF was correlated with the improved response to ICB therapy in renal cell carcinoma [17].

7.5 IMPROVING RESPONSE TO ICB THERAPY

J.J. Ishizuka et al. recently conducted a pooled in vivo CRISPR screen to identify genes expressed by the B16 transplantable melanoma model that, when deleted, confer sensitivity to immunotherapy [18]. This screen identified a number of genes with the potential to modify the response to endogenous RNA species, including Adarl, which encodes an adenosine deaminase that binds to and limits the sensing of endogenous double-stranded RNA (dsRNA).

J.J. Ishizuka et al. showed that Adarl functions as a checkpoint that limits anti-tumor immunity by preventing the sensing of endogenous dsRNA. Loss of function of Adarl improves responses to PD-1 blockade and overcomes common mechanisms of resistance to immunotherapy. B16 tumor cells lacking Adarl are hereafter termed Adarl-null tumors. Adarl-null B16 tumors in wild-type, immunocompetent animals were profoundly sensitized to anti-PD-1 antibody treatment.

7.5.1 Loss of ADAR1 Sensitizes Tumors to IFNs

Adarl-null tumors were significantly more sensitive than control tumors to T-cell killing. Enhanced T-cell killing of Adarl-null tumors could result either from improved T-cell cytotoxicity or from an increased sensitivity to secreted effector cytokines such as IFNυ. Adarl-null cells showed significant inhibition of viability and increased apoptosis, when stimulated with IFNβ or IFNυ, but not TNF. Thus, the sensing of either type I or type II IFNs is sufficient to cause growth arrest and apoptosis in Adarl-deficient tumor cells. Cytokine and chemokine genes such as Ifnbl, I16, Ccl5, Cxcl9, and Cxcl10 were upregulated in Adarl-null cells following IFN stimulation. Thus, Adarl-null tumor cells increase expression of antiviral cytokines and chemokines following exposure to IFN.

Deletion of Adarl overcomes several common mechanisms of acquired resistance to immunotherapy and causes inflammatory repolarization of the TME (tumor microenvironment) even in the absence of *CD*8+ T-cell recognition of MHC I on tumors.

7.5.2 Discussion

Loss of function of Adarl in tumor cells removes a checkpoint that normally restrains sensing of IFN-inducible dsRNA, leading to heightened IFN sensitivity mediated by MDA5 and PKR (protein kinase R), respectively. This dual mechanism increases the response to immune checkpoint blockade and overcomes resistance to immunotherapy.

A dsRNA-dependent PKR is a protein induced by IFN-α and INFβ. This serine-threonine kinase suppresses protein translation and contributes to the inhibition of viral replication [3]:122.

MDA5 (melanoma differentiation-associated 5) senses dsRNA. This protein serves as viral sensors by binding to viral RNAs using an RNA helicase-like domain in their carboxy terminal. Sensing of viral RNAs activates signaling by MDA5 that leads to type I interferon production appropriate for defense against viral infection [3]:102.

E7.1 T. HONJO

In 1942, Tasuku Honjo was born in Kyoto. In 1966, he graduated from the Department of Medicine at Kyoto University, where he also attained a doctoral degree in 1975, and then in 1982, joined as a professor. He proved that PD-1 binds to PD-L1 on the surface of antigen-presenting cells, inhibiting the immune response by T-cells. Furthermore, he proved that most cancer cells express PD-L1 on their surfaces. He developed checkpoint blockade with anti-PD-1 antibody nivolumab for cancer immunotherapy, providing the fourth cancer therapy. In 2018, he was awarded the Nobel Prize in Physiology and Medicine together with James P. Allison.

E7.2 J.P. ALLISON

James P. Allison was born in Alice, Texas in the United States in 1948. He studied at the University of Texas in Austin, and he received his PhD there in 1973. Since 2012, he has been a professor at the University of Texas MD Anderson Cancer Center in Houston.

He realized the potential of releasing our immunity by blockade of inhibitory checkpoint and thereby unleashing the immune cells to attack cancers. He developed the new therapy by checkpoint blockade for treating cancer patients. He was awarded the Nobel Prize in Physiology and Medicine together with T. Honjo in 2018.

E7.3 MELANOMA

Most cancers have characteristics that reflect their origin. For example, the cells of a basal-cell carcinoma, derived from a keratinocyte stem cell in the skin, generally continue to synthesize cytokeratin intermediate filaments, whereas the cells of a melanoma, derived from a pigment cell in the skin, will often (but not always) continue to make pigment granules. Cancers originating from different cell types are, in general, very different diseases. Basal-cell carcinomas of the skin, for example, are only locally invasive and rarely metastasize, whereas melanomas can become much more malignant and often form metastases. Basal-cell carcinomas are cured by surgery or local irradiation, whereas malignant melanomas, once they have metastasized widely, are usually fatal [1]:1092.

E7.4 TUMOR MUTATIONAL BURDEN

Tumor mutational burden (TMB) is defined as the number of somatic mutations per megabase of interrogated genomic sequence. TMB is believed to be a key driver in the generation of immunogenic neopeptides displayed on MHC on the tumor cell surface, influencing the patient's response to ICB therapies. Tumor-specific neoantigens arise from somatic mutations, and can play a pivotal role in tumor-specific T-cell-mediated, anti-tumor immunity after inhibition of checkpoint signals.

TMB may be a predictive biomarker of the tumor response to ICB therapies in several cancer types. The most robust initial responses to ICB therapies were observed in melanoma and NSCLC which typically have high mutation burden owing to the mutagenic effects of ultraviolet light and tobacco smoke, respectively. Subsequently, significant associations between high TMB and the response to ICB therapies were reported in other solid tumor types [19].

8 CAR T-Cell Therapy

8.1 CAR (CHIMERIC ANTIGEN RECEPTOR)

T-cells do not have a function to pursue an antigen target the way an antibody does. By obtaining a function to pursue a specific antigen target, T-cells can strengthen their function over natural T-cells. T-cells with strengthened function need not recognize an antigen using MHC class I-peptide complex because they have the function of an antigen-specific antibody, also attacking cancer lacking the expression of MHC class I-peptide complex. The CAR T-cell was developed to create a new T-cell with this strengthened function. The CAR T-cell has a receptor with the function of a cancer-specific antibody targeting an antigen on the cancer's surface. The new type of receptor is known as a chimeric antigen receptor (CAR): a chimera is a creature with a lion's head, a goat's body, and a snake's tail in Greek mythology (Figure 8.1).

8.2 CAR T-CELL

The CAR T-cell was used to target CD19 as a cancer rejection antigen in treating ALL (acute lymphocytic leukemia), which is an aggressive cancer of malignantly transformed B-cells. CAR has an extracellular domain and an intracellular domain. The extracellular domain is an antibody recognizing human CD19, and the intracellular domain has three ITAMs (immunoreceptor tyrosine-based activation motif) (Figure E8.3) from the ζ chain of the T-cell receptor CD3 complex (Figure 8.2) [3]:266. The ITAM is fused with a co-stimulation domain from 4–1BB, a member of the TNF receptor superfamily (Figure 8.3).

CD8 T-cells with anti-CD19 CAR were found to be effective in achieving clinical remissions of ALL patients, involving side effects necessitating therapy with IVIG [3]:723.

8.3 PRODUCTION PROCESS OF CAR T-CELL

As shown in Figure 8.4, a retroviral vector with a genome (RNA) carrying a CAR transgene introduces a gene into a cell's nucleus by transfection. Retroviral particles are produced from gene encoded CAR in nucleus of transfected cell and genomically integrate with T-cells activated with anti-CD28 monoclonal antibody (mAb), producing CAR T-cells with CAR on the surfaces. A packaging cell line, shown in Figure 8.4, is a method of packaging the engineered retroviral genome into retroviral particles. The cell line is transfected with a retroviral genome encoding retroviral gene, producing retroviral gene products [20].

Figure 8.5 shows the process whereby a retrovirus enters into a cell and performs gene integration in the nucleus, and new retroviral particles are produced. In a cell, mRNA is transformed into DNA via reverse transcription. Usually, transcription is

DOI: 10.1201/9781003780663-8

FIGURE 8.1 Chimera.

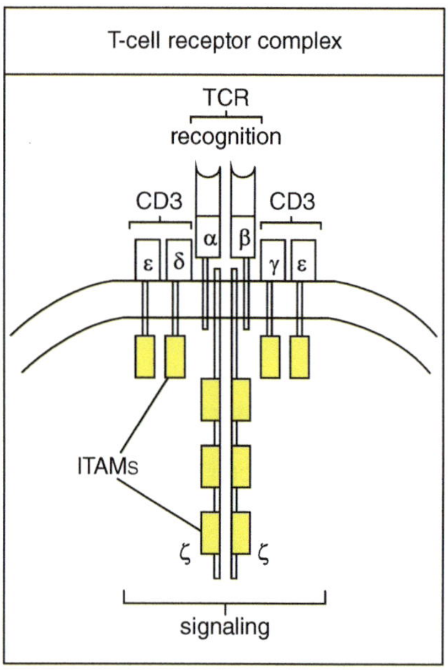

FIGURE 8.2 T-cell receptor complex [3]:266.

carried out from DNA to mRNA. In the case of a retrovirus, transcription is reversely carried out, therefore the word "retro" is used (retro meaning reverse). After adsorption of the retroviral particle, the retroviral genomic RNA passes into the cytoplasm, and is reverse transcribed into DNA. The DNA gains entry into the nucleus and

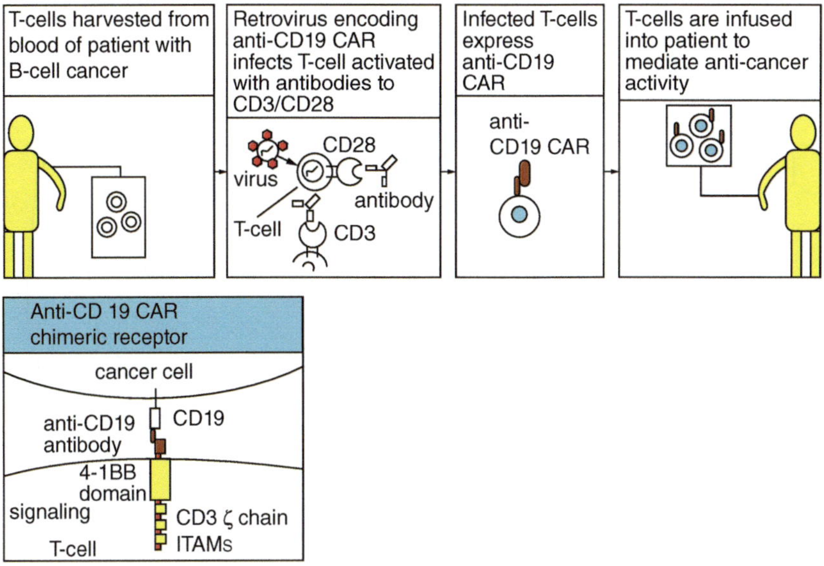

FIGURE 8.3 Chimeric antigen receptor (CAR) in T-cells [3]:724.

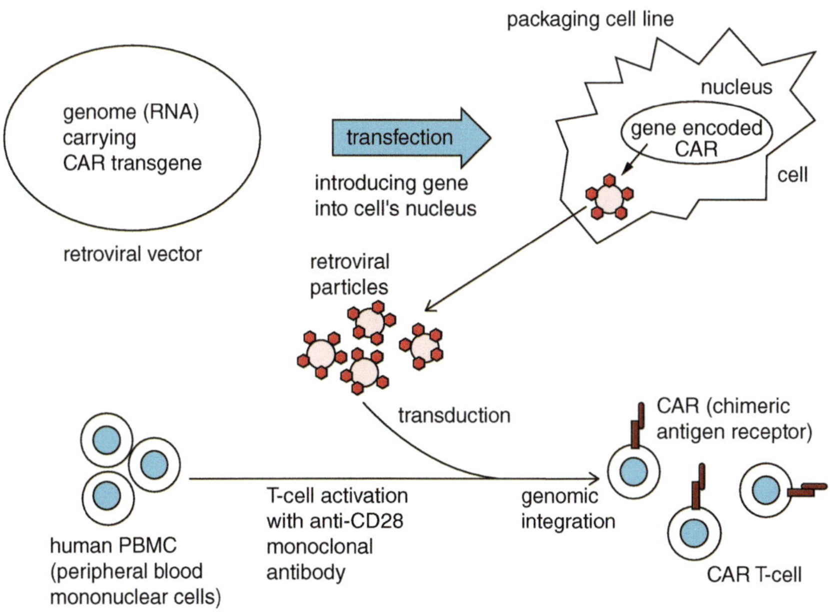

FIGURE 8.4 The process of making CAR T-cells [20].

as a proviral circle integrates randomly into the host cell chromosomal DNA. The integrated provirus acts as typical chromosomal DNA in that it is inherited by both daughter cells whenever the cell divides. The provirus is also transcribed and the retroviral genes are translated, using the cell's normal machinery. The assembly of

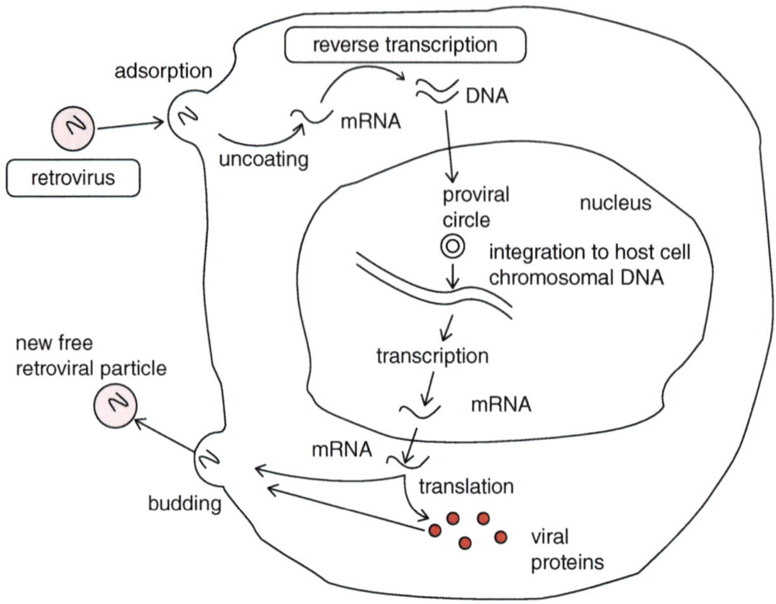

FIGURE 8.5 Retroviral life cycle [21].

a new retroviral particle completes the life cycle. The genomic retroviral RNA transcript accompanies the retroviral gene products and buds off to form a new free retroviral particle [21].

8.4 THE IMPORTANCE OF THE CAR T-CELL

The advent of CAR T-cells for leukemia and lymphoma is noteworthy in future prospects. It is most important that the CAR T-cell is the first form of gene transfer (E8.2) therapy approved by the FDA. In 2017, the FDA approved tisagenlecleucel (Kymriah), a CAR T-cell treatment for ALL [22]. CAR T-cell engineering will advance for applying these therapeutic approaches more broadly to solid cancers such as breast, lung, and head and neck cancers.

8.5 NEW CAR CONSTRUCTION AND TARGETS IN CLINICAL TRIALS

Efforts to improve upon the early CAR T-cell therapy aim to further improve response rate and duration of remission, target new disease subtypes, and decrease the substantial toxicity frequently experienced with many of the standard designs. Potential mechanisms of CAR T-cell failure include an immunosuppressive tumor microenvironment, tumor antigen escape (e.g., CD19 loss), and CAR T-cell exhaustion and decreased persistence. To this end, novel CAR designs have recently been translated from the preclinical setting to early-phase "first-in-human" studies.

CARs targeting CD22 receptors on B-cells, as well as bispecific CD19/CD22 CARs have been developed. Several trials investigating bispecific CAR T-cells are currently underway. Preliminary data from a phase 1 dose-escalation study investigating a dual anti-CD19 and anti-CD22 bivalent loop CAR design were reported by Hossain et al. at the 2018 American Society of Hematology (ASH) annual meeting. Although patient numbers were small, this study demonstrated proof of principle for efficacy and safety of bispecific CARs. Out of five patients with DLBCL (diffuse large B-cell lymphoma), there were two CRs (complete responses) (with one ongoing at 6 months) and one PR (partial response). There were no occurrences of grade 3 or higher toxicities.

The range of diseases successfully treated by CAR T-cell therapy continues to expand, as exemplified by the promising results from trials of CARs targeting B-cell maturation antigen (BCMA) in multiple myeloma (MM). In phase 1 first-in-human trial, 11 patients with heavily pretreated relapsed or refractory multiple myeloma were treated with a "home-grown" anti-BCMA CAR (MCARH171). Preliminary results were reported by Mailankody et al. at the 2018 annual meeting, which demonstrated a 64% ORR (overall response rate) and a median duration of response of 106 days. There were no dose-limiting toxicities or cases of severe neurotoxicity, and only two patients experienced grade 3 or higher CRS (cytokine release syndrome) [23].

8.6 THE NEXT GENERATION OF CAR T-CELLS

The next generation of CAR T-cells is developed as armored CARs, which include those that constitutively express CD40 ligand, secretable cytokine IL-18, or secretable PD1-blocking single-chain variable fragment (scFVs). IL-18 is an immune-stimulating cytokine that is predominantly secreted by monocytes, macrophages, and dendritic cells.

It augments the cytotoxic activity of NK and CD8 T-cells and induces CD4 T-cells to secrete IFN-γ, which stimulates nearby macrophages and antigen-presenting dendritic cells.

CD40 ligand (CD40L) is a transmembrane receptor that is expressed in activated T-cells. It stimulates a potent antitumor response by binding to CD40 on nearby B-cells, macrophages, and dendritic cells. Modified T-cells, which constitutively express CD40L, demonstrate enhanced proliferation, increased secretion of IFN-γ and GM-CSF, and downregulation of exhaustion markers such as PD–1. When co-cultured with human CLL (chronic lymphocytic leukemia) cells, upregulation of costimulatory, adhesion, human leukocyte antigen (HLA), and pro-apoptotic (Fas) receptors have been reported. Rafiq et al. report that treatment with a new CAR construct that constitutively secretes PD-1-blocking scFv molecules led to significantly improved survival [23].

8.7 GENE THERAPY

In gene therapy, therapeutic oligonucleotides have been used since 1990s. They are in the form of siRNA (short interfering RNA) or consist of only about 20 nucleotides, directed against RNA and therefore referred to as antisense RNA, which specifically targets disease-causing mRNA, cutting off the mRNA and suppressing the occurrence of genetic disease by inhibiting the translation. One example is nusinersen (Spinraza), which was approved in the US in 2016 for the treatment of spinal muscular atrophy [22]. Furthermore, using AAV (adeno-associated virus) vectors for delivery of high doses of AAV, patients suffering from retinal dystrophy were successfully treated with the product Luxturna (AAV2 expressing human RPE65), approved by the FDA in 2017 [24, 25]. This product, Luxturna, is able to deliver a normal copy of the RPE65 gene to retinal cells for the treatment of biallelic RPE65 mutation-associated retinal dystrophy, i.e. a blinding disease [25]. The word "biallelic" means the following: pertaining to both alleles of a single gene (paternal and maternal). For example, biallelic mutation carriers have a mutation (not necessarily the same mutation) in both copies of a particular gene (a paternal and a maternal mutation) (NCI Dictionaries).

In the last three decades, extensive research and new technologies have contributed to the advent of gene therapy as a promising platform to treat many hereditary human diseases, leading to remarkable successes and treatment options for diseases that were previously thought untreatable. Nevertheless, in viable gene therapy, several challenges remain, including understanding and preventing genotoxicity from integrating vectors, improving gene transfer, preventing the immune responses that limit the therapeutic efficacy of systematically administered vectors, and overcoming manufacturing and regulatory hurdles [24].

E8.1 US FDA (FOOD AND DRUG ADMINISTRATION)

In 1906, the Federal Food and Drugs Act was passed by the effort of Dr. Harvey Washington Wiley. The act is referred to as the "Wiley Act." In 1986, the FDA (Food and Drug Administration) was founded to keep people's health by guaranteeing the safety of food, drugs, and products used by consumers. A pharmaceutical manufacturer must receive approval from the FDA before initiating clinical trials.

E8.2 GENE TRANSFER

The DNA sequence that already exists is the raw material of evolution. In this sense, no gene is ever entirely new. Innovation can occur in several ways (Figure E8.1) as follows.

1. Intragenic mutation: an existing gene can be randomly modified by changes in its DNA sequence, via various types of errors that occur mainly in the process of DNA duplication.
2. Gene duplication: an existing gene can be accidentally duplicated, creating a pair of initially identical genes within a single cell.
3. DNA segment shuffling: two or more existing genes can break and rejoin to make a hybrid gene consisting of DNA segments that originally belonged to separate genes.

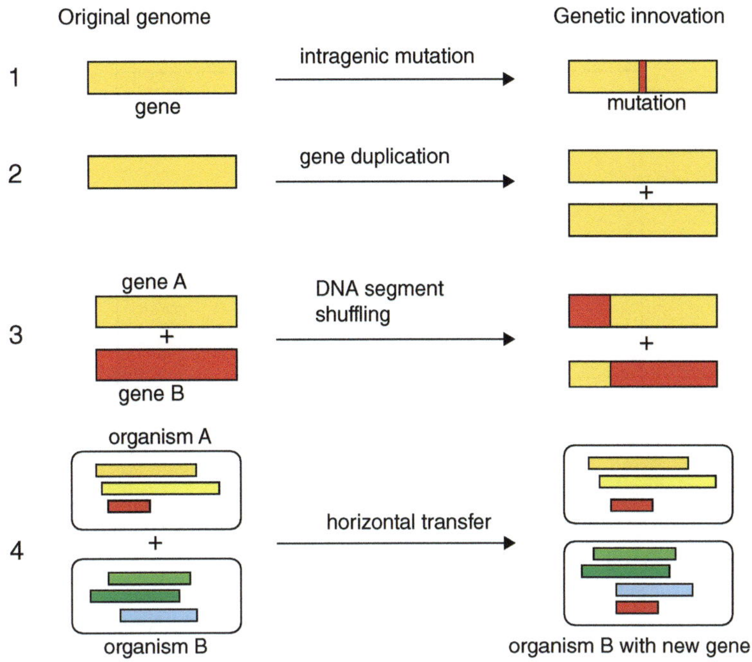

FIGURE E8.1 Horizontal transfer occurs when two different cells enter into a symbiotic association [1]:17.

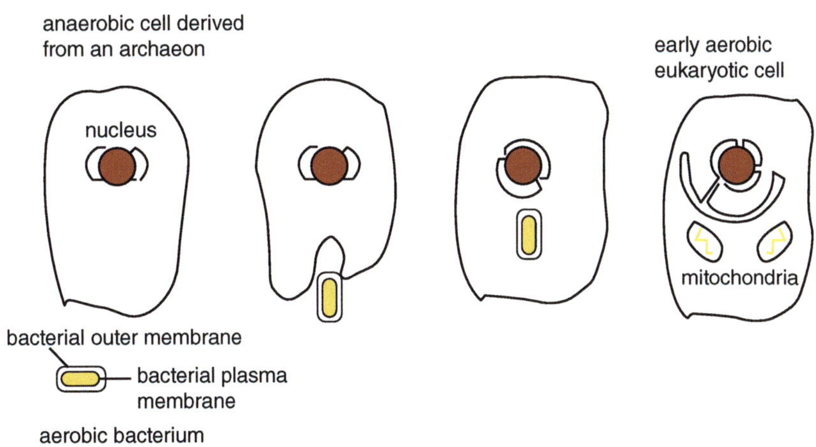

FIGURE E8.2 Mitochondria initiating a symbiotic relationship [1]:27.

4. Horizontal (intercellular) transfer: a piece of DNA can be transferred from the genome of one cell to that of another – even to that of another species. This process is in contrast with the usual vertical transfer of genetic

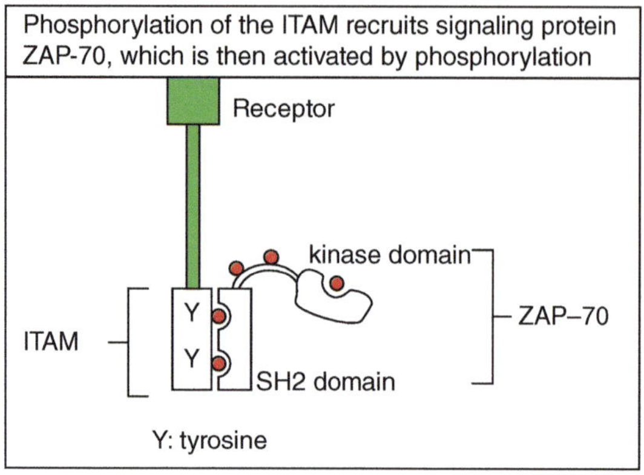

FIGURE E8.3 ITAM and signaling protein ZAP-70 [3]:267.

information from parent to progeny. This gene transfer is used in the process of CAR T-cell production [1]:16–17.

It is now generally accepted that mitochondria originated from free-living oxygen-metabolizing (aerobic) bacteria that were engulfed by an ancestral cell that could otherwise make no such use of oxygen (that is, was anaerobic). Escaping digestion, these bacteria evolved in symbiosis with the engulfing cell and its progeny, receiving shelter and nourishment in return for the power generation they performed for their hosts. This partnership between a primitive anaerobic predator cell and an aerobic bacterial cell is thought to have been established about 1.5 billion years ago when Earth's atmosphere first became rich in oxygen (Figure E8.2) [1]:25–26.

E8.3 ITAM (IMMUNORECEPTOR TYROSINE-BASED ACTIVATION MOTIF)

Tyrosine phosphorylation within cytoplasmic regions called ITAMs (immunoreceptor tyrosine-based activation motif) in the CD3ϵ, γ, δ, and ζ chains initiates signaling from the T-cell receptor. CD3γ, δ, and ϵ each contain one ITAM, and each ζ chain contains three, giving the T-cell receptor a total of 10 ITAMs (Figure 8.2).

Each ITAM contains two tyrosines that become phosphorylated by a specific signaling protein: tyrosine kinase ZAP-70 (ζ-chain-associated protein), when the receptor binds its ligand, providing sites for recruitment of the SH2 domains of the signaling proteins (Figure E8.3). The two tyrosines of the ITAM make it particularly efficient in recruiting signaling proteins that contain two tandem SH2 domains. When two tyrosines of the ITAM are phosphorylated, tandem SH2 domain-containing proteins such as ZAP-70 are recruited. This leads to the phosphorylation of ZAP-70, an essential step in the activation of these kinases. The signaling protein kinase ZAP-70 carries the T-cell activation signal onward [3]:266–270.

9 ALS (Amyotrophic Lateral Sclerosis)

The neurodegenerative disease amyotrophic lateral sclerosis (ALS, or Lou Gehrig's disease) is associated with an accumulation and abnormal assembly of neurofilaments in motor neuron cell bodies and in the axon, leading to defective axonal transportation. The degradation of the axons leads to muscle weakness and atrophy, which is usually fatal.

ALS is a fatal motor neuron disease (MND) characterized by degenerative changes in both upper and lower motor neurons. Onset typically occurs in late middle life and presents as a relentlessly progressive muscle atrophy and weakness, with the effects on respiratory muscles limiting survival to 2–4 years after disease onset in most cases. The incidence of this disease in the USA in 2014 was 5 per 100,000 people. The worldwide standardized incidence of ALS was 1.68 per 100,000 persons. Only 5% of patients survive 20 years from the time of diagnosis. Up to 10% of ALS-affected individuals have at least one other affected family member, defined as having familial ALS (fALS); almost all of these cases have been found to be inherited in an autosomal dominant manner. The remaining 90–95% of ALS cases occur in people with no prior family history; these individuals are said to have sporadic ALS (sALS).

The most common fALS-linked genes are SOD1 (superoxide dismutasel), C90RF72, FUS (fused in sarcoma), and TARDBP. In sALS cases, however, diagnostic advancements have only helped in explaining a fraction of cases, with the etiology remaining unexplained in over 90% of patients.

9.1 DIAGNOSIS OF ALS PATHOGENESIS

Resolution of the relationship between upper and lower motor neuron dysfunction, particularly the site of disease onset, is critical in the understanding of ALS pathogenesis. Advances in neurophysiological techniques for assessing upper motor neuron function, particularly the development of the threshold tracking transcranial magnetic stimulation (TMS) technique, have suggested the importance of cortical hyperexcitability in ALS pathogenesis. Consequently, novel treatment approaches aimed at modulating cortical hyperexcitability, utilizing repurposed therapeutic agents, may lead to the development of effective treatment strategies for ALS. The identification of cortical hyperexcitability may also be used as a diagnostic biomarker, facilitating an earlier diagnosis of ALS and enabling recruitment into clinical trials at a stage in the disease process at which therapeutic agents are likely to be most successful. TMS excites neurons in the underlying motor cortex with motor-evoked potentials

DOI: 10.1201/9781003780663-9

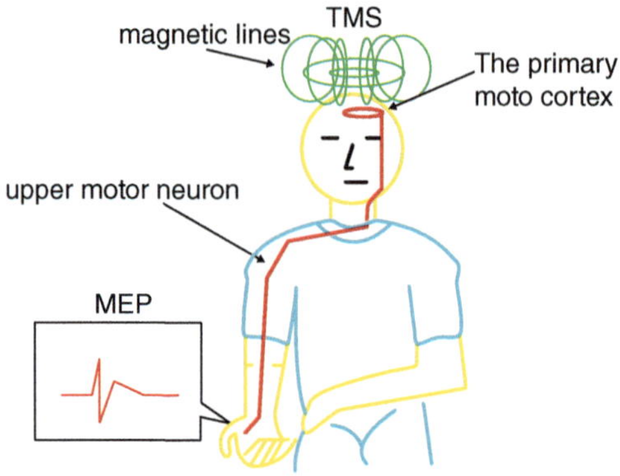

FIGURE 9.1 Measure of cortical excitability by TMS [26].

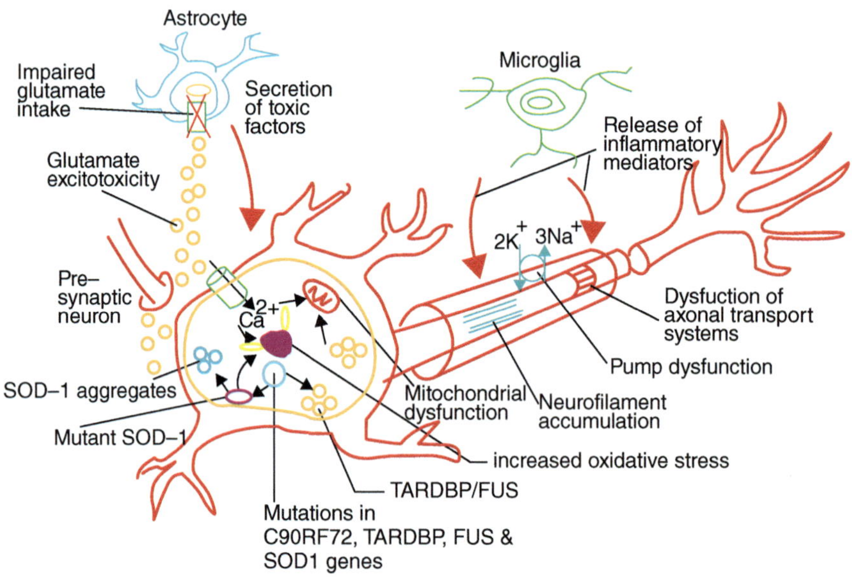

FIGURE 9.2 ALS appears to be mediated by a complex interaction between molecular and genetic pathways [26].

(MEP) recorded over the contralateral muscle. It is a sensitive measure of cortical excitability (Figure 9.1).

9.2 PHYSIOLOGICAL DYSFUNCTION IN ALS

It appears that a complex interaction between genetic and environmental factors underlies the development of ALS, leading to dysfunction of critical molecular pathways and ultimately neurodegeneration (Figure 9.2). Reduced uptake of glutamate from the synaptic cleft, leading to glutamate excitotoxicity, is mediated by dysfunction of the astrocytic excitatory amino acid transporter2 (EAAT2). The resulting glutamate-induced excitotoxicity induces neurodegeneration through activation of Ca^{2+}-dependent enzymatic pathways. Mutations in the C9ORF72, TDP-43, and fused in sarcoma (FUS) genes result in dysregulated RNA metabolism leading to abnormalities of translation and formation of intracellular neuronal aggregates. Mutations in the superoxide dismutase 1 (SOD-1) gene increase oxidative stress, induce mitochondrial dysfunction, and lead to intracellular aggregates and defective axonal transportation. Separately, microglia activation results in secretion of proinflammatory cytokines and neurotoxicity.

9.3 CLINICAL INSIGHTS

The identification of concomitant upper and lower motor neuron signs as a pathognomonic feature of ALS suggests the importance of cortical dysfunction in ALS pathogenesis. Eisen and colleagues [27] proposed the "dying forward" hypothesis, which postulated that hyperexcitability of descending corticomotoneuronal tracts mediated motor neuron degeneration via an anterograde transsynaptic glutamate excitotoxic process. A number of specific clinical observations supported this dying forward mechanism, including (1) muscles with a sparse corticomotoneuronal supply (extraocular and sphincter muscles) were relatively spared in ALS; (2) pure lower motor neuron phenotypes of ALS were infrequent; (3) absence of ALS in the animal world, a finding ascribed to developmental differences in corticomotoneuronal projections onto the spinal motor neurons across species; (4) asymmetric and focal motor deficits evident in the early stages of ALS could be explained by the complex anatomical association of corticomotoneuronal–anterior horn cell synaptic relationships.

The observation of dissociated muscle atrophy (wasting) as a specific clinical feature of ALS has further underscored the importance of cortical dysfunction in ALS pathogenesis. The split-hand phenomenon refers to preferential wasting of the thenar group of intrinsic hand muscles, namely abductor pollicis brevis (APB) and first dorsal interosseous (FDI) (Figure 9.3). This dissociated pattern of muscle atrophy is specific for ALS, and is not evident in ALS neuromuscular mimicking disorders such as Kennedy disease or autoimmune motor neuropathy. Cortical hyperexcitability

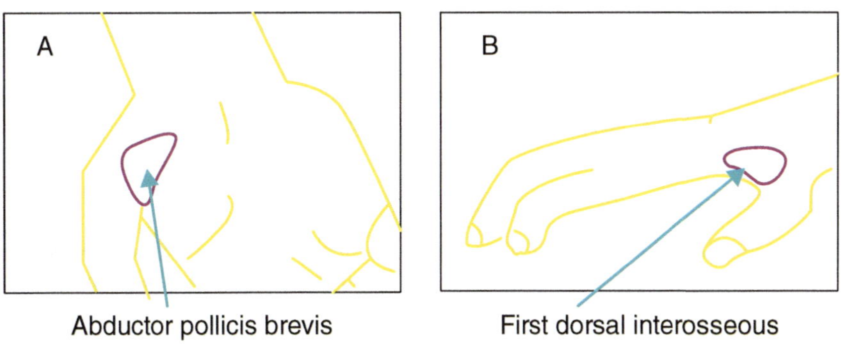

FIGURE 9.3 Dissociated muscle atrophy is a specific feature of ALS [26].

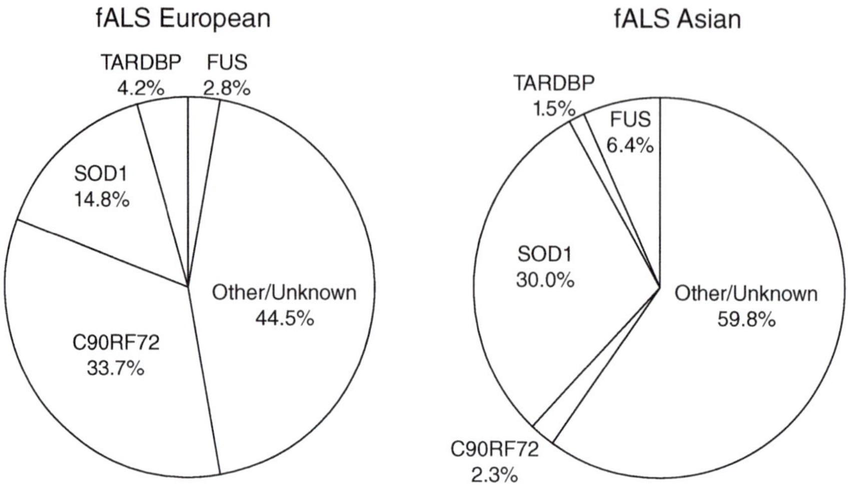

FIGURE 9.4 Proportion of ALS [28].

appears to underlie the development of the dissociated pattern of muscle wasting in ALS. In addition, a split-hand plus phenomenon, referring to preferential weakness of the APB muscle when compared to the flexor pollicis brevis, was also reported as a specific clinical feature of ALS mediated by cortical hyperexcitability. More recently, split leg and split elbow were also reported as specific clinical features of ALS and related to cortical dysfunction.

9.4 ALS-ASSOCIATED GENES

9.4.1 SOD1

The SOD1 gene (encoding superoxide dismutase 1[Cu/Zn]) was the first to be associated with ALS 1993. SOD1 encodes a 153 amino acid metalloenzyme, one of three superoxide dismutase enzymes found in humans (Figure 9.4).

The protein binds copper and zinc and forms an extremely stable homodimer. SOD1 dimers reside in the cytosol and the intermembrane space of mitochondria, providing an important antioxidant defense mechanism by catalyzing the production of oxygen and hydrogen peroxide from the superoxide species produced during cellular respiration. A recent metanalysis found that pathogenic variants in SOD1 account for approximately 15–30% of fALS and fewer than 2% of sALS.

9.4.2 TDP-43 Encoded by TARDBP Gene

In 2006, a shift in the understanding of ALS pathogenesis occurred with the discovery that the main component of the ubiquitinated protein aggregates found in ALS patients was TARDNA-binding protein 43 (TDP43). The aggregation of TDP-43 in ubiquitin-positive cytoplasmic neuronal inclusions in the brain and spinal cord is now considered a pathological hallmark of ALS.

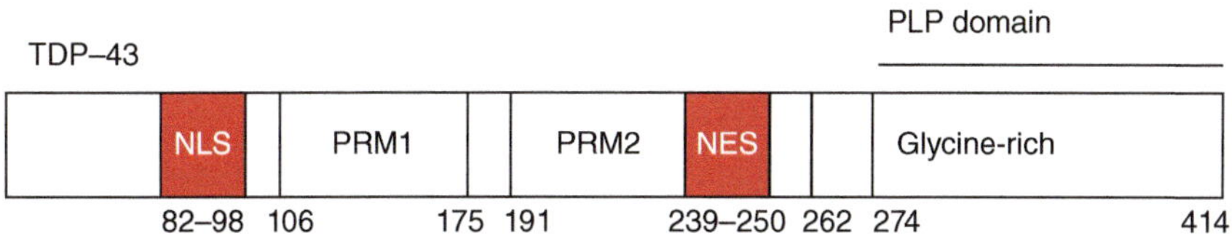

FIGURE 9.5 Domain structure of TDP-43 [28].

TARDNA-binding protein 43 (TDP-43) is a DNA/RNA binding protein composed of 414 amino acids, encoded by the TARDBP gene. Although usually concentrated in the nucleus, TDP-43 contains both a nuclear localization signal and a nuclear export signal (Figure 9.5) and shuttles back and forth between the nucleus and cytoplasm. In Figure 9.5, numbers refer to amino acid positions. NLS is nuclear localization signal. NES is nuclear export signal. PRM1 and PRM2 are RNA recognition motifs. Prion-like protein (PLP) domain spans from amino acids 274–414. ALS-associated mutations are clustered in the glycine-rich region.

In 2008, dominant mutations in the TARDBP gene were identified as a primary cause of ALS, providing evidence that aberrant TDP-43 could be causative of neurodegeneration [28].

9.5 DISEASE MECHANISM

9.5.1 SOD1 Disease Mechanism

Variations in SOD1 have been associated with a decrease in enzyme activity of 50–80%, leading to an early proposition that the disease was conferred through a loss of dismutase activity. Mutation-induced conformational and functional changes of SOD1 have been proposed to confer toxicity via interactions with many proteins and through several mechanisms. These include excitotoxicity, oxidative stress via upregulation of reactive oxygen species, endoplasmic reticulum stress, mitochondrial dysfunction, and prion-like propagation.

As in other neurodegenerative diseases involving protein aggregation, debate surrounds whether the soluble or aggregated forms of protein are responsible for exerting toxicity.

9.5.2 TDP-43 Disease Mechanism

Further histological studies have confirmed that TDP-43 is present in the cytoplasmic aggregates of the majority of ALS patients including sporadic cases (sALS) without pathogenic variants in the TARDBP gene, and in those with C90RF72 hexanucleotide repeat expansions.

Both the loss and overexpression of TDP-43 are causative of disease, indicating the importance of tightly controlled regulation of this protein. There is increasing evidence that ALS may be caused by aberrant TDP-43 regulation. Expression of TDP-43 is autoregulated through a feedback mechanism by which the protein binds

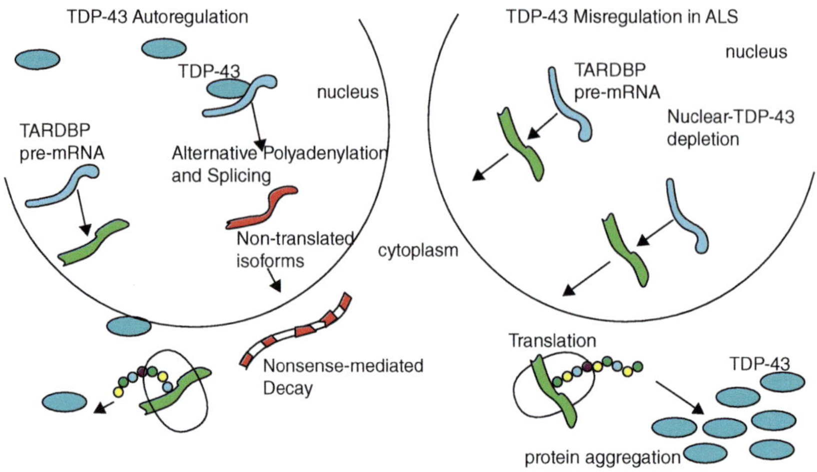

FIGURE 9.6 Schematic of TDP-43 autoregulation and misregulation in ALS [28].

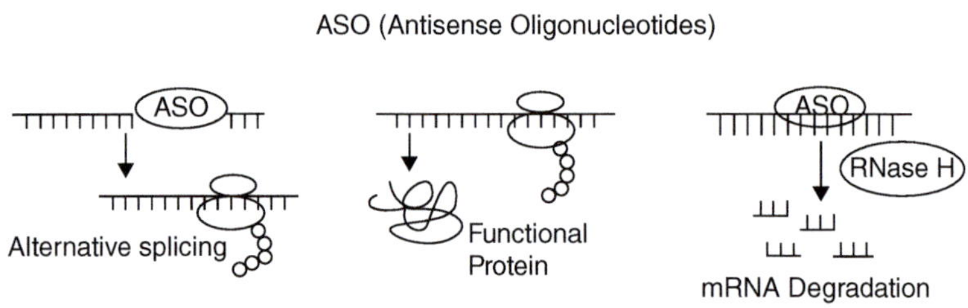

FIGURE 9.7 Non-viral strategy using ASO to induce alternate splicing or RNase H-mediated degradation [29].

to a region within the 3'UTR of its own pre-mRNA when in nuclear excess, triggering the use of alternative polyadenylation signals and splicing events that result in mRNA transcripts that are degraded rather than translated.

The depletion of nuclear-TDP-43 (Figure 9.6 right) is thought to result in the continuous upregulation of TDP-43 synthesis. Excess TDP-43 in cytoplasm may result in the formation of inclusion bodies leading to cellular dysfunction.

9.6 THERAPY FOR ALS

9.6.1 ASO (Antisense Oligonucleotides)

ASOs targeting RNA were used in therapeutic approaches to treat fALS. These short, single-stranded nucleic acids can affect the gene by modifying, splicing, or inducing mRNA degradation by activating ribonuclease H (RNase H) (Figure 9.7).

In 2006, Smith et al. showed a good distribution of ASO within the central nervous system, effective suppression of rat SOD1 mRNA, and decreased levels of

Gene Therapy	Agent	Phase	Date
SOD1	Tofersen (BIIB067)	I I/II III	2013 2020 current
C90RF72	BIIB078	I	2021
FUS	Jacifusen (ION363)	III	current
ATXN2	BIIB100	I	current

FIGURE 9.8 ASO therapy clinical trials in ALS [30].

SOD1 protein in nervous tissues. It was also associated with slowing the course of the disease and prolonged survival. Phase 1 study confirmed the safety and good tolerability of intrathecal ASOs in humans. Phase 1–2 studies on tofersen (BIIB067) showed promising results; this is an ASO that reduces SOD1 protein synthesis by mediating RNase H dependent SOD1 mRNA degradation (Figure 9.8).

9.6.2 AAV (Adeno-Associated Virus Vector)

The AAVrh10 serotype, which, due to its transduction into the central nervous system, can effectively deliver microRNA there, which, by binding SOD1 mRNA, would thus reduce the production of mutant proteins in patients with this form of ALS (Figure 9.9). Therefore, it would improve the survival and function of motor neurons, resulting in therapeutic benefits for people with SOD1-associated ALS. The study of Biferi et al. (2017) used AAVrh10 vectors to mediate hSOD1 pre-mRNA exon skipping. This led to the production of a premature termination codon, and as a result, increased survival in the tested mice and prevented the deterioration of neuromuscular function in them.

Currently, the safety, tolerability, and efficacy of AAVrh10 intrathecal administration of anti-SOD1 microRNA to patients with SOD1 ALS mutations is being tested in the multi-center clinical trial APB–102.

9.6.3 Genome Editing

An approach recently tested for the treatment of both genetic and non-genetic disorders is genome editing. Its technologies include zinc finger nuclease, transcription activator-like effector nuclease, and Cas9 nuclease associated with CRISPR (clustered regularly interspaced short palindromic repeats).

In order to use CRISPR-Cas9, a small guide RNA (sgRNA) is first constructed close to the sequence of a PAM (proto-spacer adjacent motif). As a result, in the next step, Cas nuclease is able to bind to the target DNA sequence, which is recognized by the sgRNA. In effect, the pattern leads to a double strand break (DSB) in the cell's DNA, inducting base insertion or deletions to cause a frameshift. These breaks are in turn joined by two different repair processes: non-homologous end joining (NHEJ) and homologous recombination (HR). The first of these techniques, NHEJ, is a fast

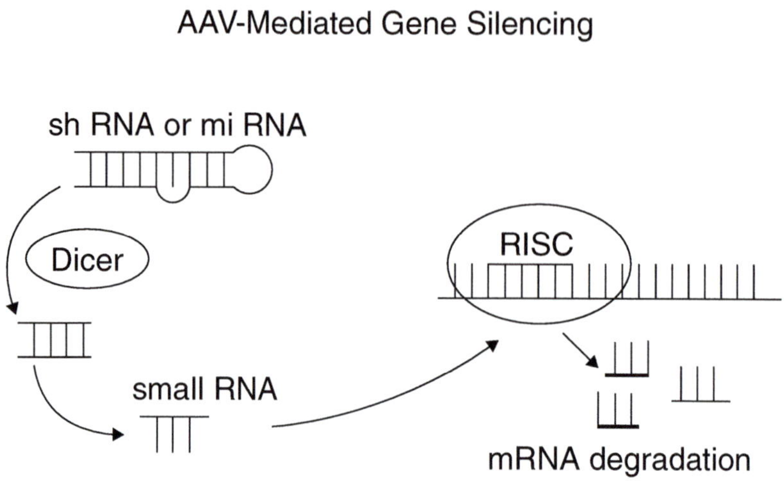

FIGURE 9.9 AAV-mediated gene silencing through RNA interference [29].

but error-prone process that often leads to insertions and deletions at the breakpoint, thereby knocking out the gene encoding the protein. In contrast, HR induces insertion or replaces DNA fragments with a donor template, while being relatively error-free (Figure 9.10).

In the study by Pribadi et al., the CRISPR-Cas9 system was used to completely remove the large repeat expansion mutation within C90RF72 in induced patient-derived pluripotent stem cells (iPSC), which have the ability to differentiate into other types of cells, such as neurons. This work opens up further possibilities for the use of CRISPR-Cas9 in cutting out the mutant C90RF72 repeat expression in ALS therapy (2016).

E9.1 TMS (TRANSCRANIAL MAGNETIC STIMULATION)

TMS is a procedure that uses a magnetic field to stimulate nerve cells in the brain for assessing upper motor neuron function of patient with ALS. TMS is called a noninvasive procedure because it is done without using surgery or cutting the skin.

During a TMS procedure, an electromagnetic coil is placed against the scalp of the head. This coil delivers magnetic pulses that stimulate nerve cells in the region of the brain. It is thought to activate regions of the brain that have decreased activity during depression [31].

E9.2 TDP-43

TDP-43 remains a dominant protein of interest to understand disease pathogenesis. TDP-43 was identified as a primary component of ubiquitinated and hyper-phosphorylated cytosolic aggregates observed from post-mortem tissue of patients with ALS. This pathological phenomenon is considered a hallmark of ALS because it is

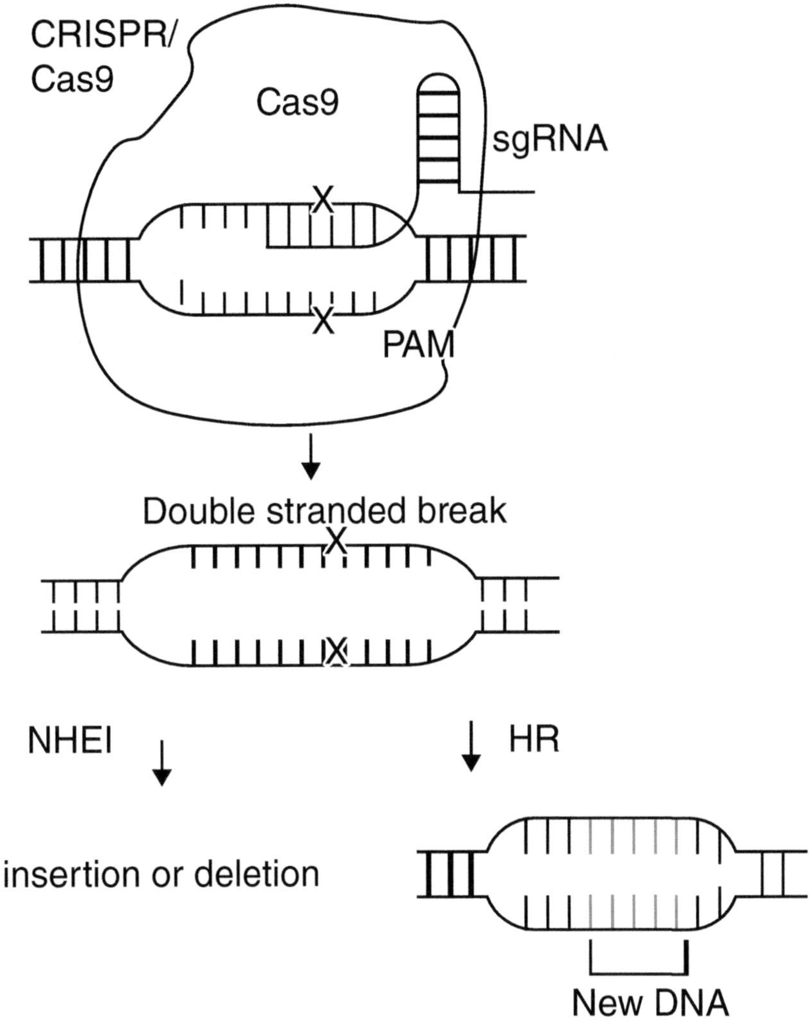

FIGURE 9.10 Technique of CRISPR-Cas9 genome editing system [30]. TMS is also used to improve symptoms of major depression. The FDA approved TMS for obsessive compulsive disorder (OCD) and migraines, and to help stop smoking when standard treatments have not worked well. Research continues into other potential uses for TMS, including epilepsy.

observed in approximately 97% of all ALS patients regardless of the mechanisms of disease onset with the notable exceptions of familial ALS (fALS) caused by mutations in Zn/Cu superoxide dismutase I (SOD I) and fused in sarcoma (FUS). Since the first report in 2008, over 50 mutations in TARDBP have been linked to ALS, further supporting TDP-43 dysfunction as a critical component in ALS. Therefore, TDP-43 dysfunction provides common ground in an otherwise convoluted disease, thus gaining notoriety and attention from researchers aiming to uncover the mechanisms causing TDP-43 aggregation.

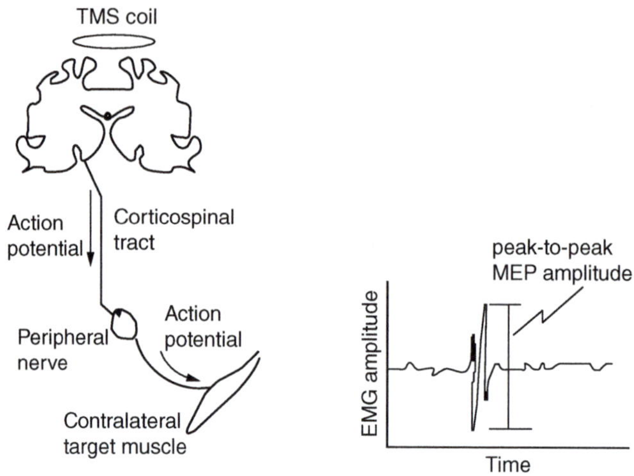

FIGURE E9.1 TMS-MEP [33].

It is important to note that mutations in TARDBP can also cause frontotemporal lobar dementia (FTLD), which itself shares some clinical parallels with ALS and displays TDP-43 pathology in 45% of cases [32]. Deletion or overexpression of TDP-43 causes toxicity or cell death in cell and animal models.

E9.3 MEP (MOTOR-EVOKED POTENTIAL)

Transcranial magnetic stimulation (TMS) applied over the motor cortex activates interneurons oriented parallel to the brain surface. This coil placement leads to a trans-synaptic activation of pyramidal cells evoking descending volleys in the corticospinal tract. Motor neuron activation in response to corticospinal volleys leads to a brief contraction in the muscle termed a motor-evoked potential (MEP), which is recorded through electromyography (EMG) surface electrodes applied over the muscle belly (Figure E9.1). The MEP can be used to estimate excitability of the corticospinal tract, while the silent period which follows the MEP can be used to estimate corticospinal inhibition [33].

E9.4 MOTOR NEURON

ALS was originally characterized in 1869 by Jean-Martin Charcot; however, it was not until the diagnosis of the famous New York Yankee in 1939, Lou Gehrig, that the disease become more well known. Motor neuron diseases (MND) are neurological disorders characterized by a loss of varying motor neurons resulting in decreased physical capabilities.

Metabolic dysfunction is heavily represented in MNDs. Studies currently reveal that altered metabolism could either be the cause or the result of disease pathogenesis.

Although motor neurons are a primary focus when investigating MNDs, similar metabolic dysfunction has been discovered in skeletal muscle tissue, including localization at the neuromuscular junction (NMJ). Therefore, understanding dysregulated metabolism in both the central nervous system (CNS) and periphery is necessary for diagnostic and therapeutic intervention.

The CNS houses highly metabolic cell types requiring copious amounts of energy to maintain physiological functions. The intricate communication between cell types in order to maintain metabolic balance has been extensively reviewed. The brain consumes about 20% of the body's total energy in order to maintain proper functions. Defects in any of the complexes have been shown to result in decreased energy production in neurodegenerative disease.

Connections formed at the NMJ are important for muscle function and contraction. An action potential is delivered to synapses allowing for the release of acetylcholine (Ach); Ach can then traverse across the NMJ and bind to the Ach receptor on the muscle fiber, enabling a contraction downstream. Metabolic balance at the NMJ is necessary for maintaining communication between neuronal and skeletal muscle cells. It is known that mitochondria ATP production and neurotransmitter release at the NMJ is necessary to maintain a stable synaptic connection [34].

Appendix A: Healthcare in the UK

Among 7 developed countries, the United Kingdom (NK) was evaluated as the second rank in comprehensive healthcare, and as the first rank in efficiency and effect, according to the healthcare report "Mirror, Mirror on the Wall" in 2010 [35].

In the NHS (National Health Service), public health service is not free access, but functions are strictly shared. The primary care is provided to patients by a GP (general practitioner). Citizens are demanded to register with a GP for general medical services. The GP plays the role of a gatekeeper. A patient cannot receive a higher health service without the approval of the GP, except an emergency medicine.

The second health service is provided by hospitals, giving special medical care, mental disorder care, and emergency care. The third health service is provided by college hospitals.

In the United Kingdom, only a person registered to the General Medical Council (GMC) can provide health service. When a person passes the graduation examination at a medical college (6 years) approved by the GMC, the person can obtain a medical license automatically. After graduation, according to the foundation program, the person ought to receive clinical training for two years. After clinical training, the person who hopes to become a GP receives training in emergency medicine, pediatrics, and psychiatry for 2–3 years, and then works as GP. The person who hopes to become a specialist doctor furthermore receives training to obtain a CCT (Certificate of Completion of Training).

Bibliography

1. B. Alberts et al., *Molecular Biology of the Cell,* Sixth Edition, Garland Science, 2015.
2. B. Alberts et al., *Essential Cell Biology,* Fourth Edition, Garland Science, 2014.
3. K. Murphy et al., *Immunobiology,* Ninth Edition, Garland Science, 2017.
4. R. Koch, Kyushu University Medical Library, https://www.lib.kyushu-u.ac.jp/hp_db_f/igaku/expl/koch.html.
5. M. Hukuda, "Shibasaburo Kitasato: Pioneer of Treatment for Infectious Diseases," https://www.nippon.com/en/japan-topics/b07213/.
6. I. Lee, "Introduction of mRNA Vaccine Against COVID-19," https://www.cas.org/ja/resource/blog/covidmrna-vaccine.
7. K.S. Park et al., "Non-Viral COVID-19 Vaccine Delivery Systems," *Advanced Drug Delivery Reviews,* 169, 137–151, February 2021.
8. "Institute Pasteur," https://zaidan.pasteur.jp/en/institute/index.html.
9. T. Shioyama, *Principle and Application of Sensor* (in Japanese), Morikita Shuppan, 2002.
10. Wikipedia, "Philadelphia Chromosome," https://en.wikipedia.org/wiki/Philadelphia_chromosome.
11. H. Kobayashi, "A New Type of Anti-Cancer Therapy by Near-Infrared Photoimmunotherapy (in Japanese)," *Pharmacology,* 76(3), 172–176, 2016.
12. IBM Blog Japan, "Toward Realization of Ultimate Therapy Without Metastatic Tumor and Relapse of Tumor," https://www.ibm.com/blogs/think/jp-ja/mugendai-8462-interview-nci-cancer-treatments-follow-up/.
13. H. Wakiyama et al., "Near Infrared Photoimmunotherapy of Cancer; Possible Clinical Applications," *Nanophotonics,* 10(12), 3135–3151, 2021.
14. K. Sato, *A Mechanism of Cancer Cell Cytotoxicity of Near Infrared Photoimmunotherapy,* Nippon Laser Igakkaishi, Advance Publication released online June 27, 2020.
15. S. Sakaguchi, "Regulatory T Cells and Human Diseases," *Annual Review of Immunology,* 38, 54–66, 2020.
16. T. Honjo et al., "Current Issues and Perspectives in PD-1 Blockade Cancer Immunotherapy," *International Journal of Clinical Oncology,* 25, 790–800, 2020.
17. C.M. Fares et al., "Mechanisms of Resistance to Immune Checkpoint Blockade: Why Does Checkpoint Inhibitor Immunotherapy Not Work for All Patients?" *American Society of Clinical Oncology Educational Book,* 39, 147–164, May 17, 2019.
18. J.J. Ishizuka et al., "Loss of ADAR1 in Tumours Overcomes Resistance to Immune Checkpoint Blockade," *Nature,* 565(7737), 43–48, 2019.
19. D. Sha et al., "Tumor Mutational Burden (TMB) as a Predictive Biomarker in Solid Tumors," *Cancer Discovery,* 10(12), 1808–1825, December 2020.
20. R.A. Morgan and B. Boyerinas, "Genomic Modification of T Cells," *Biomedicines,* 4, 9, 2016.
21. J. Price, "Retroviruses and the Study of Cell Lineage," *Development,* 101, 409–419, 1987.
22. E. Smith and P. Blomberg, "Gene Therapy-from Idea to Reality," *Lakartidningen,* 114, EWYL, December 19, 2017.
23. D. Sermer and R. Brentjens, "CAR T-Cell Therapy: Full Speed Ahead," *Hematological Oncology,* 37(S1), 95–100, 2019.

24. J.P. Tremblay and A. Annoni, "Three Decades of Clinical Gene Therapy: From Experimental Technologies to Viable Treatments," *Molecular Therapy,* 29(2), 411–412, February 2021.
25. G.A. Rodrigues et al., "Pharmaceutical Development of AAV Based Gene Therapy Products for the Eye," *Pharmaceutical Research,* 36, Article number 29, 2019.
26. M.A.J. van den Bos et al., "Pathophysiology and Diagnosis of ALS: Insights from Advances in Neurophysiological Techniques," *International Journal of Molecular Sciences,* 20, 2818, 2019.
27. A. Eisen et al., "Amyotrophic Lateral Sclerosis (ALS): A Phylogenetic Disease of the Corticomotoneuron?" *Muscle Nerve,* 15, 219–224, 1992.
28. R. Mejzini et al., "ALS Genetics, Mechanism, and Therapeutics: Where Are We Now?" *Frontiers in Neuroscience,* 13, 1310, 2019.
29. D.A. Amado and B.L. Davidson, "Gene Therapy for ALS: A Review," *Molecular Therapy,* 29(12), 3345–3353, 2021.
30. J. Lejman et al., "Gene Therapy in ALS and SMA: Advances, Challenges and Perspectives," *International Journal of Molecular Sciences,* 24, 1130, 2023.
31. "Transcranial Magnetic Stimulation," https://www.mayoclinic.org/test-procedures/transcranial-magnetic-stimulation/about/pac-20384625.
32. T.R. Suk et al., "The Role of TDP-43 Mislocalization in Amyotrophic Lateral Sclerosis," *Molecular Neurodegeneration,* 15, 45, 2020.
33. "Motor Evoked Potential," https://www.sciencedirect.com/topics/neuroscience/motor-evoked-potential.
34. A.E. Kalyuzhny and N. Brustovetsky, "Altered Metabolism in Motor Neuron Diseases: Mechanism and Potential Therapeutic Target," *Cells,* 12(11), 1536, June 2, 2023, https://doi.org/10.3390/cells12111536.
35. Mirror, "Mirror on the Wall: How the Performance of the U.S. Health Care System Compares Internationally," June 23, 2010 (Report), Commonwealth Fund, 2010 Update, http://www.commonwealthfund.org/Publications/Fund-Reports/2010/un/Mirror-Mirror-Update.aspx?page=all.
36. CDC.gov., "Understanding mRNA CVID-19 Vaccines," https://www.cdc.gov/coronavirus/2019-ncov/vaccines/differentvaccines/mrna.html.

Index

P

R

S

T

U

V

W

Y

Z

For Product Safety Concerns and Information please contact our EU representative GPSR@taylorandfrancis.com Taylor & Francis Verlag GmbH, Kaufingerstraße 24, 80331 München, Germany

Batch number: 10726519

Printed by Printforce, the Netherlands